IN SEARCH OF PLANET VULCAN

The Ghost in Newton's Clockwork Universe

Statue of U. J. J. Le Verrier at the entrance to the Paris Observatory, with Le Verrier's signature superimposed. (Photograph courtesy David Graham. Image processing by Julian Baum.)

IN SEARCH OF PLANET VULCAN

The Ghost in Newton's Clockwork Universe

RICHARD BAUM
and
WILLIAM SHEEHAN

PLENUM TRADE • NEW YORK AND LONDON

Library of Congress Cataloging-in-Publication Data

Baum, Richard, 1930-
   In search of planet Vulcan : the ghost in Newton's clockwork
universe / Richard Baum and William Sheehan.
     p.   cm.
  Includes bibliographical references and index.
  ISBN 0-306-45567-6
  1. Vulcan (Hypothetical planet) 2. Planets--Miscellanea.
3. Planets--Research.   I. Sheehan, William, 1954- . II. Title.
QB605.2.B38 1997
523.4--DC21   96-45674
                                       CIP

ISBN 0-306-45567-6

© 1997 Richard Baum and William Sheehan
Plenum Press is a Division of Plenum Publishing Corporation
233 Spring Street, New York, N.Y. 10013-1578
http://www.plenum.com

10 9 8 7 6 5 4 3 2 1

Printed in the United States of America

For Audrey and Julian
—R. B.

To my parents Bernard and Joyce Sheehan
—W. S.

Preface

To the people of the late 19th century, Vulcan was real. It was a planet. It had theoretical credibility and had actually been seen. Even textbooks accorded it a chapter. When the spell was finally broken in the early 20th century, it dropped to the floor of history, where it has gathered dust, redundant, unwanted. It was something best forgotten, an embarrassment. In simple terms, that is all true.

The truth, however, is of two Frenchmen. Urbain Jean Joseph Le Verrier, the imperial mathematician who discovered planets from his desk, and Edmond Modeste Lescarbault, an obscure country physician who one sunny afternoon in March 1859 fulfilled his dream of other worlds by discovering Vulcan. This is the planet that grew out of Le Verrier's attempt to extend planetary theory to the limits of gravitational mechanics, specifically his struggle with the errant motion of Mercury, the innermost planet.

Again, that is only part of the story. For the whole affair derived substance from Le Verrier's reluctance to question the integrity of the ancien régime of the lawgiver himself, Sir Isaac Newton. This plunged astronomy into a long period of intellectual trench warfare. Only when astronomers were prepared to modify the foundations of the ancien régime did they finally break the impasse. This was a conceptual shift of historical significance, a watershed. It led to a quickening of thought about gravitational theory and a restructuring of celestial mechanics. Vulcan, then, is the surf that marks the edge and the advance of understanding. It is something thrown up on the frontier of knowledge that signified a feature of the physical world we had yet to recognize. And yet it is something else.

"N.Y. astronomer discovers a planet circling the Sun closer than Mercury" proclaimed a press headline in 1979. "Is the sun surrounded by a ring?" queried another. The gap between 1859 and 1979 closed and the name Vulcan was again on many lips. As if to compound the hope, that same year an unidentified stellarlike object was photographed near the eclipsed Sun. No longer the ghost in the Newtonian clockwork, Vulcan still haunts the imagination.

This feeling is confirmed if one visits Orgères-en-Beauce, a small community in rural France some 20 miles or so north of Orléans, déparêtment of Eure et Loire. There, in March 1859, from a small cupola overlooking flat open country, Lescarbault made his celebrated observation. Later in the year, an imperious knock on his door obliged him to relive that moment. Soon the world knew every detail. Today Orgères-en-Beauce acknowledges those events with a road named in honor of Lescarbault. A tangible reminder of the day the town pulsed and Monsieur le docteur began his movement into history.

Here then is the story behind the complex science of Vulcan, a tale that highlights the logical and serendipitous, if not irrational, nature of the scientific pursuit. It is rich in the eccentricities of human character, of astronomers far from the popular ideal. Other than Lescarbault, there is the autocratic U. J. J. Le Verrier, the mathematician who essentially created Vulcan, and James Craig (Tubby) Watson, who made the most credible (but disputed) observations of the planet at the July 1878 eclipse and who, near the end of his life, was supervising the construction of an underground observatory for the express purpose of vindicating his observations. Then, too, there is his nemesis, C. H. F. Peters, an ex-soldier of fortune who, having fled Sicily, landed in Turkey with so little money he had to choose between a supper and a cigar; he elected the latter. Years later, he was found on the threshold of his college home where he had collapsed and died after a night's observing, a half-burned cigar in his fingers. A thorough-going skeptic, Peters ridiculed the quest for Vulcan as the vain search for "Le Verrier's mythical birds." The stories of these men are joined not by similarities of taste or character nor even by continuous and intimate transaction (Peters and Watson were clearly bitter enemies), but by a passionate pursuit often bordering on obsession—the search for Vulcan. It may be called, in emulation of Kepler's celebrated warfare with Mars, the Le Verrier Campaign in the war of Mercury's secular motion, i.e., the attempt to solve the problem of the planet's errant motion within the context of Newtonian gravitational mechanics. Rather than adopt an analytical approach, the story is told in its wider context, from faint stirrings in prehistory, through the pageant of discovery epitomized in Newton's majestic conception, to denouement in the American West on a hot dusty day in July, 1878, and finale in the Trojan Vulcans, a lesser-known episode that takes the story beyond its traditional limit.

That the story should be in scale was the primary concern from the outset. Now it is told, that is still our hope.

Richard Baum and William Sheehan

Acknowledgments

This book had its genesis in "Le Verrier and the Lost Planet," a short article published in 1981, following Richard Baum's involvement the previous year in the television series "Arthur C. Clarke's Mysterious World." During the intervening years of sporadic work hundreds of people and numerous institutions have helped in countless ways. They are thanked, particularly the specialists whose works helped us to fill gaps in our knowledge. Such publications feature in the Select Bibliography. Material primary to the text appears in the Notes and References.

The following are thanked in more particular terms. Dorothy Schaumberg from the Mary Lea Shane Archives of the Lick Observatory, who supplied many of the illustrations used in the book, besides granting permission to quote from unpublished letters. Peter Hingley and Mary Chibnall of the Royal Astronomical Society responded promptly as ever to an unending stream of requests. Nor can the scholarly help of the late Dr. E. W. Maddison, a former librarian of the Society, be overlooked. Mrs. Sheila Edwards and her staff at the Royal Society were always helpful, especially in obscure matters. Understanding of C. H. F. Peters benefited from the enthusiasm of Frank Lorenz, Hamilton and

Kirkland Colleges, New York; while the late Dr. Joseph Ashbrook pointed toward Vulcan's prehistory. The reconstruction of Separation, Wyoming, in 1878 owes much to Dr. John A. Eddy, who visited the site in 1968 and 1973.

Permission to use the portraits of Asaph Hall and Albert Einstein is by courtesy of the Yerkes Observatory. The University of Wisconsin is thanked for material from the Comstock papers. The Astronomical Society of the Pacific is also thanked for its generous permission.

A word of gratitude here will acknowledge the help, but scarcely discharge the obligation owed to the Mayor of Orgères-en-Beauce for information about Edmond Modeste Lescarbault. Monique Favre, Conservateur of the Municipal Library, Chateaudun, graciously provided a register of Lescarbault papers held there.

The help of Adam Perkins, Archivist of the Royal Greenwich Observatory, Cambridge, and of Ingrid Howard, Librarian, Royal Greenwich Observatory is also acknowledged. Dr. Steven Dick, U.S. Naval Observatory, kindly sent a fascinating extract from his still unpublished history of the U.S. Naval Observatory. Dorrit Hoffleit, Yale University, supplied information about Edward Herrick and his searches for intramercurial planets.

The headline "N.Y. astronomer discovers planet orbiting sun closer than Mercury," is a reminder of the late Charles F. Capen, Jr. His interest in the subject is best summed up by his 1970 remark, "Somehow, I hope there really is a planet Vulcan."

Special thanks go to David Graham, whose photograph of the statue of U. J. J. Le Verrier at the entrance to the Paris Observatory provided an opportunity for a unique illustration.

The chapters on the Neptune affair, as the late Colin Ronan called it, benefit directly from memorable discussions with Dr. Robert W. Smith, Hubble Space Telescope historian, then of Merseyside Museums, England, now of the National Air and Space Museum, Washington D.C., and Johns Hopkins University. His extensive knowledge encourages the hope of a comprehensive history of the subject at some future time. Dennis Rawlins is another noted authority whose work is acknowledged with respect.

Contemporary scholarship owes much to his meticulous research, as does the discussion of the much maligned French astronomer Pierre Charles Lemonnier. Special regard must be assigned to Dr. Allan Chapman, Wadham College, Oxford for his insights into the life and work of George Biddell Airy and William Lassell.

Professor Richard S. Westfall graciously granted permission to quote from *Never at Rest,* his magisterial biography of Sir Isaac Newton. Patrick Moore very generously allowed sight of his translation of Liais's account of the discovery of Neptune. Quotations from H. M. Harrison's *Voyager in Time and Space: The Life of John Couch Adams, Cambridge Astronomer* (1994) are thanks to Chantal Porter, Managing Editor, The Book Guild Ltd. Those from *The Herschel Chronicle* (1933) and the multivolume *The General History of Astronomy* are by courtesy of the publisher, Cambridge University Press, with special thanks to Dr. Adam Black.

Dr. Donald Osterbrock, director emeritus of the Lick Observatory, author and astronomical historian of repute, is thanked for invaluable material. His deft and incisive criticism of an early draft of the manuscript helped to improve its style and argument. Thomas A. Dobbins made several useful suggestions, notably in regard to Lescarbault's telescope. Gerard Gilligan, Alan W. Heath, Norris Hetherington, Jim Lattis, Lou Marsh, and Dr. Richard McKim were very helpful in various matters.

Final preparation of the manuscript was expedited through the expertise of Julian Baum, who is also thanked for providing the dustjacket illustration and line diagrams. His technical skill proved invaluable.

In conclusion, indebtedness is owed to Linda Greenspan Regan at Plenum, who saw in the Vulcan story what others had missed. Without her support, encouragement, and enthusiasm, the book almost certainly would have taken much longer to produce.

Contents

Introduction

"If I knew of someone who applied himself to Mercury, I believe I would be obliged to write to him in order to charitably counsel him to better spend his time," advised Michael Maestlin, sometime tutor of Johannes Kepler.[1] Part of the problem was the difficulty astronomers had in making accurate observations of the planet from which to construct a satisfactory theory of its movements. The situation improved in the 17th and 18th centuries, following the invention of the telescope and the development of more exact measuring tools. Thus 19th-century precision ought to have produced a reconciliation. Instead, it disclosed a puzzling discrepancy between the observed and theoretical motion of Mercury. When that was closed early in the 20th century, a new physics was in place. It was installed with the help of Vulcan—not the fiction of Star Trek, but a planet that was a poltergeist, the product of a young Frenchman's mathematics. The Frenchman's name was Urbain Jean Joseph Le Verrier, son of an employee of the State Property Administration in Normandy. Early in his astronomical career he determined upon a solution to the riddle. His first attempt in 1843 gave a shadowy result. However, the second time he opened his box of intellectual tools, some 16 years later, out popped Vulcan.

U. J. J. Le Verrier was one of the great theoretical astronomers of the 19th century, a colossus whose work, in the words of Norwood Russell Hanson, "has lain largely undiscovered; a suboceanic mountain beneath the scientific sea."[2] He was a hero of classical physics, of "normal science." The paradigm within which he worked was Newtonian theory, a majestic, many-jewelled clockwork that seemed, for an illusive moment, to be self-adjusting, stable, and eternal. His ambitious goal, like that of his illustrious predecessors Alexis Claude Clairaut, Leonhard Euler, Jean Le Rond d'Alembert, Joseph Louis Lagrange, and the Marquis Pierre Simon de Laplace, was to account for every nod and wobble of the solar system by the universal law of gravitation—one of the grandest aspirations ever conceived by the human mind. The Sun distorted the Moon's ellipse around the earth, so that it dilated and contracted, gyrated and precessed. Planet tugged on planet—all this occurred according to law.

Le Verrier turned to celestial mechanics in 1837 after abandoning a promising career in chemistry. Two years later he established his credentials with an essay on the stability of the solar system. It was an impressive debut, heralding "to the scientific world the appearance of a potential successor to Laplace."[3]

In September 1846, he was 35 and at the pinnacle of his profession. Four months earlier he had published an investigation into the motion of Uranus, the planet discovered by William Herschel in 1781. Uranus' wayward movements had long been a bane of mathematical astronomers, until even the theory of universal attraction itself was suspect. Le Verrier, however, rather than charge the law with insufficiency, accounted for Uranus' wanderings by invoking the gravitational influence of an unknown exterior body.[4] That body had now been discovered; Neptune, the eighth planet, had been added to our inventory of the solar system.

The discovery seemed strangely uncanny. Le Verrier had, as Camille Flammarion put it, "discovered a star with the tip of his pen, without other instrument than the strength of his calculations alone." It was an unparalleled triumph, quite literally the "Zenith of Newtonian Mechanics," in Hanson's evocative phrase.[5] Le Verrier was acclaimed, and 1846 slipped into history, checked as one of the most memorable years in the annals of astronomical science. And yet at the time of its most spectacular victory, Newton's system of celestial government faced its gravest threat. In a sense, Le Verrier had fortified a Trojan horse under the banner of Neptune.

Even before Uranus engaged his interest, Le Verrier had skirmished with the equally perplexing motions of Mercury, the innermost primary of the system, of which he had said: "No planet demands of astronomers more attention and pains, and none grants them in recompense so much anxiety or annoyance."[6] It was to be his epitaph.

As is well known, Mercury is small, moves with great speed, and has a well-defined perihelion, this being the position on its orbit where it most closely approaches the Sun. Were Mercury and the Sun alone in the solar system, the line between the perihelion and the Sun would remain forever and unalterably fixed in space, but the other planets pull on Mercury and cause this line to precess slowly forward. From an analysis of the planet's transits across the sun, Le Verrier found he could not reconcile its observed rate of precession with that predicted by Newtonian theory.

He first encountered the problem in 1843 and produced an imperfect theory, but did not press the matter. In part this was probably due to feelings of insecurity produced by his lack of prominence in the astronomical community, and partly because he doubted the accuracy of his mathematics.

Things were very different after the discovery of Neptune. Le Verrier was a man of distinction, the accomplished director of the Paris observatory. Newton's law, too, was supreme. The ingenuity of the mathematicians had risen superior to every obstacle. Although some problems yielded with difficulty, all yielded, as the motion of Uranus had done, to the gravitational formulae. There was no reason to believe the excess advance of the perihelion of Mercury would be an exception. Intractable as it seemed, it was after all only a minor flaw.

Accordingly Le Verrier embarked upon a detailed investigation of the problem, and in 1859 gave notice of the likelihood inside the orbit of Mercury of either one or more planets, or perhaps a ring of asteroids. As these intramercurial objects must frequently pass in front of the Sun, he cautioned astronomers to be vigilant of any unusual dark spot they might see in the course of their regular studies.

Of course there was nothing startling or new in the hypothesis. It merely gave credence to an idea that had been rumored since the earliest days of telescopic astronomy. There were a number of false alarms of suspicious black specks that crept slowly but perceptibly across the face of the Sun. Indeed it was Lescarbault's one great passion to catch an unknown interior planet at such a moment. This was the raison d'être of his interest in the Sun, and why on the afternoon of March 26, 1859, he just happened to see a dark round spot enter and exit its disk. "Could it be . . . ," we can almost hear him sigh, as the object disappeared from sight. Yet it is obvious from his 9-month delay in reporting the observation that but for Le Verrier and his hidden planet hypothesis, no more would have been heard of the dark spot of Orgères. Lescarbault later explained he had hoped to repeat the observation before making an announcement. But the object did not reappear. Is it not true then to say the story of Vulcan turns on expectation and coincidence and no small amount of self-delusion?

Le Verrier caused history. Lescarbault made it. Together, yet independently, they created Vulcan. To Le Verrier, it was a mathematical necessity, a cog in the celestial machinery. To Lescarbault, it was less abstract. He held up a planet for all to see and fired the popular imagination. For a shining moment, "Garibaldi and the weather ceased to interest the Parisians; and the village doctor, in his extempore observatory, and his round black spot, appropriately bearing the name of Vulcan, were the only subjects of discussion, and the only objects of learned and unlearned admiration," as 13 years earlier Neptune had been.[7] This time however, his success was ambiguous. The dignified quest that had led to the careful building up of the "Majestic Clockwork," culminating in the discovery of Neptune, unraveled into the mad pursuit of a phantom.

THE MAJESTIC CLOCKWORK

one

The Elusive
Planet of Twilight

The quest for Vulcan begins in the study of the wayward movements of the planets, and so, in a real sense, in the myth and super-
stition of prehistory when vigilant watchers of the night sky first
recognized the wandering stars (*planetae* in Greek). What were these
moving lights? What purpose did they serve? How were they able
to move with such precision across the starry heavens?

The Greeks called the planets after the gods of Olympus,
names the Romans later alchemized into those by which they are
still known—Mercury, Venus, Mars, Jupiter, and Saturn. Mercury
was always more difficult to capture than the rest, a wayward
fugitive on the margins of the skies only now and then crossing
out of the dusky twilight or predawn mists into the realm of observation. Among the Chaldean shepherds, in the translucent
skies of the near Middle East, it was Ninob, Nabou, or Nébo, the
god who alone knew how to rouse the Sun from its bed. The association invoked intelligence, if not wiliness and craft. Among the
early Greeks it was Stilbon, the Scintillating One, and "the elusive
and nimble-footed." From this came its later identification with
Hermes, fleet-footed messenger of the gods of Olympus, one and
the same with the Roman Mercury.[1]

Swift and elusive, a constant source of anxiety to the mathematical astronomer, Mercury's positions eventually furnished the critical data in the quest for Vulcan.

The apparent movements of Mercury, and indeed of the rest of the planets, are disconcertingly complex. As we now know, this is a result of their apparent paths reflecting the inextricably mixed movements of both the Earth and the planets around the Sun. At times they move more swiftly, at other times more slowly. At still other times they reverse their motion and move backward or retrograde for a period of several weeks, before resuming their eastward motion among the stars.

As inner planets, Mercury and Venus appear alternately on either side of the Sun and never wander very far from its side—Mercury remaining always within 28°, and Venus within 47° of the Sun. They may pass behind the Sun (superior conjunction) or between the Earth and Sun (inferior conjunction). Their retrograde loops occur around the time of inferior conjunction. The outer planets—Mars, Jupiter, and Saturn—make their retrograde loops around the time they lie in opposition to the Sun in the sky, a time they also reach their greatest brilliance.

The Greeks did more than register the planets as lucid points of light upon the inky backdrop of the night. They not only looked, but studied and meditated. They charted the wanderings of the planets among the constellations of the Zodiac—the constellations that lie along the *ecliptic*, the Sun's apparent path through the sky. Among various explanations for what they saw, the most interesting, by modern standards, was that of Aristarchus of Samos about 250 B.C. Aristarchus arranged all the planets in orbits around the Sun and proposed that the Earth itself was an ordinary planet. His idea was not widely supported at the time.

Instead the more comfortable geocentric view prevailed, and Aristarchus's most brilliant successors, Apollonius and Hipparchus, tinkered with an intricate ferris-wheel machinery, centered on the Earth, to account for the apparent movements of the planets. This was the famous (or infamous) system of epicycles (Figure 1). The Earth rested in the center; the Sun, Moon, and planets moved around it, each on its own large circle known as

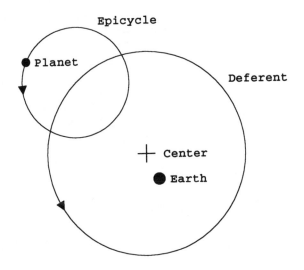

1. The epicycle and the deferent. After Aristarchus (about 310–230 B.C.) until the time of Copernicus, the epicycle was the most favored theory of planetary motion. In the general case, the planet moves uniformly in a circle known as the epicycle, the center of which is moving around a larger circle called the deferent.

the deferent, to which was attached a smaller circle, the epicycle. The movements in the deferent and epicycle occurred in opposite directions and together caused the planet to periodically swing inward toward the Earth, producing the apparent backward or retrograde motions.

This system was developed and perfected during the 2nd century A.D. by Claudius Ptolemy, a Greek astronomer who lived in Alexandria. Fairly or unfairly, Ptolemy's theory (Figure 2) has acquired the reputation of being hopelessly cumbersome, a compilation of epicycles upon epicycles. The latter accusation is patently untrue; there was never more than a single epicycle— even for Mars, a critical case. On the other hand, the theory was obviously jerry-rigged to fit the observations. In the Middle Ages, Alfonso of Castile sponsored tables of planetary motions based

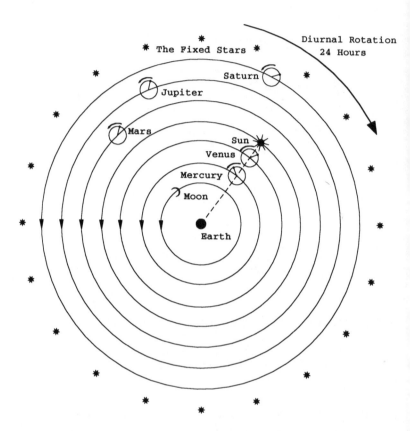

2. The Ptolemaic system. Geocentric model of the solar system described by Claudius Ptolemy (2nd century A.D.). Elaborated from ideas advanced 3 centuries earlier by the Greek astronomer Hipparchus (2nd century B.C.), it was generally accepted in the Arab and Western worlds until superseded by the heliocentric model in the 16th and 17th centuries.

on Ptolemy's theory, which was the standard for 2 centuries after his death in A.D. 1284. Alfonso is said to have complained of Ptolemy's astronomy that it was "a crank machine; that it was a pity the Creator had not taken advice."

This is probably apocryphal, and yet for Mercury, Ptolemy's construction was cumbersome enough to justify such a remark.

Here the uncertainty of the observations, rather than deficiencies of the basic theory, led to complications. Ptolemy would have done perfectly well to represent Mercury's motion with the same basic construction he devised for the other planets. Why he was tempted to do otherwise is not entirely clear. Harvard historian Owen Gingerich has proposed that since Ptolemy believed Mercury occupied the space between the Moon and the other planets and since he was legitimately forced to improvise more intricate theories to explain the Moon's baffling behavior, "he probably imagined that Mercury, as an intermediate object, also required extra complexity."[2] Mercury is almost impossible to observe in certain positions in its orbit from middle northern latitudes, particularly at its morning elongations in April, when it rises with Aries the Ram, and at its evening elongations in October, when it sets with Libra the Balance. Thus, Ptolemy incorrectly deduced that there were two points at which the apparent diameter of Mercury's epicycle would be greatest as seen from the Earth. To account for this, he constructed a hypothesis in which the planet made two close sweeps to the Earth. This double-perigee solution was quite literally a "crank" machine; it worked just like a crank—the center of the deferent moved about a little circle, thrusting the epicycle back and forth along a line. One cannot help admiring the Alexandrian astronomer's ingenuity, but obviously the little planet, which even at its best just skirted the Sun's rays, was destined to give astronomers headaches. It had already begun to do so long before Vulcan was a twinkle in their eyes.

Ptolemy's Earth-centered system continued to dominate astronomical thought until 1543, when Copernicus, a canon in the Cathedral at Frauenburg (now in Poland), published his immortal book *de Revolutionibus Orbium Caelestium*. Here, 17 centuries after Aristarchus, he revived the heliocentric or Sun-centered model. This was a decisive step. However, Copernicus was otherwise conservative and retained the machinery of epicycles. Mercury's motion again required considerable tinkering chiefly because Copernicus placed too much faith in Ptolemy's observations (indeed, he published no new observations of Mercury of his own). Thus, he noted, "this planet has ... inflicted many perplexities and labors on us in our investigation of its wanderings."[3]

More exact planetary observations were made by the Danish astronomer Tycho Brahe, who was born 3 years after Copernicus's death. For 20 years he had a magnificent observatory, the fabulous Uraniborg, on the island of Hveen, in the Sund between Copenhagen and Elsinore Castle. (All his observations were made with the naked eye, since the telescope had not yet been invented.) However, after quarreling with his subjects, he was forced to leave and eventually relocated in Prague. In 1600 he was joined by a gifted young German mathematician, Johannes Kepler (Figure 3).

3. Johannes Kepler (1571–1630). (Courtesy of the Mary Lea Shane Archives of the Lick Observatory.)

After Brahe's death a year later, Kepler assumed control of the master's observational notebooks. He used the observations of Mars, the planet to which Brahe had paid the closest attention, to trace the true shape of the orbit of that planet. It was, he found in 1605, an ellipse, with the Sun at one focus. Kepler was fortunate he had based his work on Brahe's observations of Mars, since its orbit is eccentric enough for him to have made his great discovery. Had he begun with another planet, such as Venus with its nearly circular orbit, the solution might well have eluded him.

Five parameters are necessary to define a Keplerian orbit in space. These, together with their mathematical symbols, which this book will refer to, are: the distance between the planet and the Sun, a; a measure of the departure of the elliptical orbit from the circular form, known as the eccentricity e; the inclination i, or tilt of the planet's orbit relative to the ecliptic; the longitude of the perihelion $\tilde{\omega}$, the point in its orbit where it lies closest to the Sun, thus fixing the orientation of the orbit in space[4]; and one or the other of the nodes, the ascending node Ω or the descending node $\mho$, the points where the plane of the planet's orbit intersects the ecliptic.[5] (Figure 4).

In order to calculate the position of the planet at any given time, it is necessary to know its position at some specific date. This is given by ϵ, the longitude at epoch, or by τ, the time of one of the planet's passages through the perihelion of its orbit (the point in its orbit where it lies closest to the Sun).[6]

In the Ptolemaic scheme, Mercury and Venus were set between the Earth and the Sun. This was merely a convention. It would have been equally reasonable to place their orbits beyond the Sun. On the other hand, in the Copernican theory, the ambiguity of their position was removed—Mercury and Venus were clearly interior to the orbit of the Earth.

One important implication is that Mercury and Venus can sometimes pass directly in front of the Sun, so as to appear as a moving black spot upon it. These passages, known as transits, occur at inferior conjunction—that is, when Mercury and Venus lie between the Earth and Sun. Usually, because Mercury's orbit is tilted by some 7° to the plane of the Earth's orbit and Venus' by

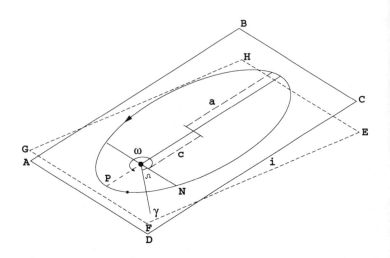

4. Orbital elements. ABCD = orbital plane of a planet or comet around the Sun. EFGH = plane of the ecliptic. i = inclination of orbital plane to ecliptic. γ = First Point of Aries. $\tilde{\omega}$ = direction of axis with respect to line of nodes. N = Ascending node. Ω = longitude of ascending node. e = the eccentricity, one of the important parameters used to define an orbit $e=c/a$.

3.5°, they miss the Sun at these times, passing above or below it. Only when inferior conjunction takes place near one of the nodes is a transit possible. As the longitude of Mercury's descending node is the same as the longitude of the Earth on May 7 and the ascending node on November 9, transits of Mercury occur only on or near these dates. For Venus, the corresponding dates are in June and December.

Kepler predicted, but did not live to see, a transit of Mercury on November 7, 1631. It was observed by only three astronomers, of whom the most notable was Pierre Gassendi, canon of the parish church of Digny, in Provence.[7] Gassendi, observing from Paris, was surprised by the small size of the planet's disk against the Sun, and first thought he was witnessing an ordinary sunspot. However, its rapid motion soon convinced him he was seeing the planet.

"I have found him. I have seen him where no one has ever seen him before!" Gassendi exulted to Wilhelm Schickard, professor of mathematics at the University of Tübingen. Although Gassendi's realization of Mercury's small size (13 arc seconds in diameter, below the threshold of naked-eye visibility) was the most important revelation to come from the 1631 transit, the event was notable in one other respect: The planet's passage across the Sun's disk provided an exacting test of the astronomical tables of the day. For centuries this planet was regarded as a singularly unrewarding object of observation. It had now been sharply defined and precisely located upon the Sun. From his measures, Gassendi calculated the transit had actually occurred ahead of the computed time by 4 h, 49 min, 30 sec; even so, Kepler's calculations were extraordinarily good by the standards of the day—his predicted position was in error by a mere 13' in longitude, and 1' 5" in latitude, compared to nearly 5° for the then standard Ptolemaic and Copernican tables.

Astronomers would later struggle to bring the unexpectedly small planet of Gassendi ever more tightly into the lock of their calculations, the timings of its transits furnishing the most accurate measures of the perfection of their efforts. For all that, it would remain intractable, a constant source of frustration and erroneous conclusion.

Much remained to be unraveled, but Gassendi's observations started a new era. His close friend Nicolas-Claude Fabri de Peiresc wrote: "This beautiful observation which you have made of Mercury's entrance onto and exit from the Sun, I regard as among the most important in many a century."[8] Astronomers had unwittingly taken the first steps on the long road to Vulcan.

two

Le Grand Newton

One of the first converts to Kepler's elliptic astronomy was a young Englishman, Jeremiah Horrocks (or Horrox). He was born in 1619 at Toxteth, then a small village near Liverpool, and was educated at Emmanuel College, Cambridge. Horrocks's goal was to correct Kepler's theories of planetary motions in order to bring them into closer agreement with observation.[1] Perhaps his most important work was in devising an accurate theory of the motion of the Moon. He also devoted much effort to understanding the motion of Venus. In October 1639, Horrocks was at Hoole, Lancashire, a rather desolate site, bordered by a morass on the east and Marton Mere and the Douglass River on the south. There he repeated the calculations of the times of the planet's transits. Kepler had predicted no such events until 1761, but Horrocks found that a transit was due to take place on November 24, 1639 (old style; the new style date is December 4, 1639). He and his friend, William Crabtree, a draper and fellow astronomer at Broughton near Manchester, succeeded in observing it—thus becoming sole witnesses of an event not to be repeated for 122 years.

Horrocks seemed destined for a great career but he died unexpectedly in January 1641, at not yet 22. Crabtree did not outlive him long; he died 3 years later at 34. Another promising youth, William Gascoigne, the inventor of the micrometer, which gave improved accuracy in celestial measurement, also died young. He perished in the battle of Marston Moor during the English Civil Wars. Robert Grant writes in his *History of Physical Astronomy:* "Amid the angry din of political commotion, the name of Horrocks was completely forgotten."[2] Fortunately, his papers survived. Some of them fell into the hands of his friend, Christopher Townley, who lived at Pendle; he shared them with another English astronomer, Jeremy Shakerley. After emigrating to Surat, India, Shakerley became sole witness to the next transit of Mercury on October 23, 1651 (old style; November 3, new style). This was possibly the first telescopic observation of a celestial phenomenon to be made from the subcontinent.[3]

The next transit of Mercury occurred May 3, 1661. It was observed by Christiaan Huygens, then in London, and Johannes Hevelius at Danzig. Kepler's tables as corrected by later astronomers such as Ismael Boulliau of France and Giambattista Riccioli of Italy "came neer the truth, and failed not many minutes."[4]

The May 3, 1661, transit happened on the same day as the coronation of Charles II, which formally closed the period when the Puritans under Oliver Cromwell ruled England following the Civil Wars and the execution of Charles I. One era had passed, another was beginning—as in politics, so in science. A new and dominating figure would appear on the scene. A month after the transit, on June 5, 1661, Isaac Newton entered Trinity College, Cambridge.

Isaac Newton was born Christmas Day 1642 (old style), the posthumous son of a yeoman farmer who had never learned to sign his own name. He lived until the age of 3 at the manor house of Woolesthorp, near Grantham, and after his mother's remarriage to wealthy Barnabas Smith, rector of nearby North Witham, remained at Woolesthorp with his grandparents Ayscough. The manor house in which Newton spent his youth still stands, a T-shaped structure of gray limestone set upon the open plains of Lincolnshire.

When he began to attend the grammar school in Grantham, Newton took up lodgings with Mr. Clark, the local apothecary. He undoubtedly passed a painfully lonely boyhood; his school fellows rejected him for obvious reasons, as he was, in the words of his early biographer William Stukeley, "commonly too cunning for them in every thing."[5] He amused himself by making lanterns, which he flew at night using kites, and other mechanical toys and sundials. Eventually, poor Clark's house was filled with sundials.

Newton's future was in question after he finished grammar school. Rev. Smith had died, and Hannah, his mother, returned to Woolesthorp. She recalled her son to help her manage the farm, but the experiment proved a disaster; Newton was not suited for "low employments," and preferred reading under the hedge to farming. The servants scoffed that he was "fit for nothing but the 'Versity." His uncle, Rev. William Ayscough, and the local schoolmaster, Mr. Stokes, shared this assessment and prevailed upon Hannah to send him to Cambridge.

As an undergraduate, Newton was aloof and isolated. His sole friend, apart from his tutor Isaac Barrow, seems to have been John Wickins. Wickins, after falling out with his chamber-fellow one day, "retired . . . into ye Walks, where he found Mr Newton solitary and dejected; Upon entering into discourse they found their cause of Retiremt ye same, & thereupon agreed to shake off their present disorderly Companions and Chum together."[6] So they did for the next 20 years. Yet the friendship with Mr. Wickins (later Rev. Wickins), left little discernible trace on Newton. As Richard Westfall notes: "Genius of Newton's order does not readily find companionship in any society in any age."[7]

Newton's studies at Cambridge included Descartes's *Geometry* and the works of Gassendi. His education was interrupted by the outbreak of the bubonic plague of 1665. In London, 17,000 people died during August of that year, and another 30,000 in September. Since cases were reported even outside London, there was no real escape. The University shut down, and Newton returned to the manor house in Woolesthorp.

The plague years, 1665–1666, have been telescoped for convenience in most accounts into one remarkable "year" (Newton's *annus mirabilus*). It saw the development of the fundamental concepts of his method of fluxions (the calculus), his investigations with the prism into the nature of colors, and his first calculations concerning gravitation. This was a time of "triumphs unequaled in the history of scientific invention."[8] Derek Whiteside, editor of Newton's mathematical papers, has noted that papers from this period may "throb with energy and imagination," but they also "convey the claustrophobic air of a man completely wrapped up in himself, whose only real contact with the external world was through his books."[9]

As an old man, Newton recalled this period, telling his early biographer Stukeley that "the notion of gravitation . . . was occasion'd by the fall of an apple, as he sat in a contemplative mood."[10] A similar account, also of late authority, was recorded by Newton's nephew John Conduitt:

> [W]hilst he was musing in a garden it came into his thought that the power of gravity (wch brought an apple from the tree to the ground) was not limited to a certain distance from the earth but that this power must extend much farther than was usually thought. Why not as high as the moon said he to himself & if so that must influence her motion & perhaps retain her in her orbit, whereupon he fell a calculating what would be the effect of that supposition.[11]

Newton knew that a stone when whirled rapidly around at the end of a string would appear to pull on the string with considerable force. Descartes described this force as an "endeavour to recede from the center." The Dutch physicist Christiaan Huygens called it a "centripetal force." The same center-fleeing force would be expected for any body moving in a circular path, such as the Moon in its orbit around the Earth and the planets in their orbits around the Sun. What was it, Newton asked, that played the role of the string and kept the Moon and planets from flying off into space?

Newton's reflection on the falling apple led him to wonder whether gravity might provide this counteracting force. The apple fell 20 feet from the tree to the Earth. Would it fall just as well if the tree were 200 miles in height or even 200,000 miles, the distance from the Earth to the Moon? Newton thought so. At the greater distance of the Moon, the gravitational pull of the Earth would be weaker, and the rate of falling would be less—but it would fall just the same. In the absence of any force, the Moon would, according to the law of inertia, travel along a straight line. But the pull of gravity would cause the Moon to fall continually toward the Earth. The rate it travels forward along its orbit and at which it falls toward the Earth are balanced. Thus, it never actually reaches the ground.

Newton also suspected the force of gravity falls off according to an inverse-square law. In other words, double the distance and the force falls off to 1/4, triple it and it falls off to 1/9, and so on. He attempted a rough back-of-envelope calculation, comparing the force of gravity on the Moon with that of a stone falling to the Earth. Since the distance to the Moon is 60 times the radius of the Earth, this ratio should be 1/3600—i.e., the Moon ought to fall as far in a minute as a stone would fall in a second. He found the results "answered pretty nearly," but did nothing more to follow up at the time.

With the passing of the Great Plague, Newton returned to Trinity and threw himself into other researches, mainly mathematics and optics. Among other things, he invented the reflecting telescope, which he presented to the Royal Society in 1669. He was also absorbed in studies of alchemy and theology. (It has, in fact, been estimated that he spent at least twice as much time working on these subjects as he did on mathematics and physics.)

Meanwhile, gravitation had begun to interest others—notably, Christopher Wren, Savilian professor of astronomy at Oxford and architect of St. Paul's Cathedral, and the versatile experimentalist Robert Hooke. Both were leaders of the Invisible College for the Promoting of Physico-Mathematical Experimental Learning, chartered in 1662 as the Royal Society for the Improvement of Natural Knowledge. By 1679, they had convinced themselves that the motions of the Moon and planets were due to an attractive force. They knew this force must vary according to the inverse-square law, but neither was a good enough mathematician to take the problem further.

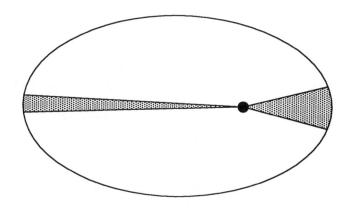

5. Kepler's law of areas. Every planet revolves so that the line joining it to the Sun sweeps over equal areas in equal intervals of time.

Hooke and Newton had not been on the best of terms for a number of years—among other things, Hooke criticized one of Newton's papers on optics. However, they renewed contact late that year. Hooke questioned Newton: What path would a body follow when attracted by an inverse-square force?[12] Newton turned the question around and, beginning with an elliptical orbit, attempted to calculate the law of force. He showed that an attractive force directed toward the center implied Kepler's law of areas (Figure 5), a profound result that later became the basis of his general method for determining orbital motion.[13] He next demonstrated that an elliptic orbit around an attracting body situated in one focus implied an inverse-square law of attraction.[14] Yet, he did not inform Hooke of his results, and again did nothing more to follow them up.

A comet appeared in late 1680—or actually, as was believed by most astronomers at the time (including Newton himself), two comets, one in the morning sky before sunrise in November and another at the end of the year in the evening sky after sunset. Still it did not yet occur to Newton to apply to it the principles of orbital motion he had established for planets around the Sun.

Nothing more was done until January 1684, when Wren, Hooke, and Edmond Halley discussed the problem of planetary motions at the Royal Society in London. At 27, Halley was the youngest of the three and already distinguished in astronomy. While still a student at Oxford, he had taken a year's leave of absence (1676–1677) to make a hazardous expedition to remote St. Helena (where Napoleon was later exiled) to draw up a catalog of southern stars. He had also observed on November 7, 1677, a transit of Mercury—only the fourth transit observed since the invention of the telescope. Wren, Hooke, and Halley all struggled with the problem of formulating the planetary motions in terms of a law of force, but the exact solution remained elusive.

In August, Halley visited Cambridge, and while there laid the problem before Newton. According to the account written by the English mathematician Abraham DeMoivre,

> The Dr asked him what he thought the Curve would be that would be described by the Planets supposing the force of attraction towards the Sun to be reciprocal to the square of their distance from it. Sr Isaac replied immediately that it would be an Ellipsis, the Doctor struck with joy & amazement asked him how he knew it, why saith he I have calculated it, whereupon Dr Halley asked him for his calculation without any farther delay. Sr Isaac looked among his papers but could not find it, but he promised him to renew it, & then to send it to him.[15]

Newton did so, and in November Halley received a nine-page treatise: *De motu corporum in gyrum* (On the Motion of Bodies in an Orbit). Impressive as Halley found it, it was no more than a preliminary draft; Newton immediately began reworking and expanding it. The majestic problem of the celestial motions had seized him, and seized him irresistibly—it would prove to be the turning point of his life.

In the waning months of 1684, Newton began investigating centripetal forces as they determined orbital motion. As he later recalled, at almost the exact moment James Stuart was proclaimed King James II of England (February 1685), Newton arrived at a new and still more powerful generalization, that of an attraction "arising from the universal nature of matter." His gravitational force between two bodies was proportional to the amount of matter they contained (their mass) and became weaker with increasing distance between them according to the inverse-square law. The outline of this powerful new principle can already be discerned in his first revision of *de Motu:*

> If in any position of the planets their common centre of gravity is computed, this either falls in the body of the Sun or will always be close to it. By reason of the deviation of the Sun from the centre of gravity, the centripetal force does not always tend to that immobile centre, and hence the planets neither move exactly in ellipses nor revolve twice in the same orbit. There are as many orbits of a planet as it has revolutions, as in the motion of the Moon, and the orbit of any one planet depends on the combined motion of all the planets, not to mention the action of all these on each other. But to consider simultaneously all these causes of motion and to define these motions by exact laws admitting of easy calculation exceeds, if I am not mistaken, the force of any human mind.[16]

To work out the consequences of this breathtaking conception, the universal attraction of every body in the universe for every other body, was the immense task Newton began in April 1685. He planned to transform the suggestive but still sketchy *de Motu* into his immortal masterpiece, the *Philosophiae naturalis principia mathematica* (Mathematical Principles of Natural Philosophy), always referred to as the *Principia.* He spent 18 months in near total seclusion and concentrated effort in his rooms in Trinity. "So intent, so serious upon his Studies," writes an eyewitness from the time, "yt he eat very sparingly, nay, oftimes he has forget to eat at all, so yt going into his Chamber, I have found his Mess untouch'd of wch when I have reminded him, [he] would reply, Have I; & then making to ye Table, would eat a bit or two standing. . . . When he has sometimes taken a Turn or two [in the garden], [he] has made a sudden stand, turn'd himself about, run up ye Stairs, like another Archimedes, [and] with an *eureka*, fall[en] to write on his Desk standing, without giving himself the Leasure to draw a Chair to sit down."[17]

In 1687, Halley, as editor of the *Principia,* husbanded it at his own expense through the press. He was the first to grasp the magnitude of Newton's achievement. In a Latin poem prefixed to the first edition, he paid the author a sublime tribute:

> O mortal men,
> Arise! And, casting off your earthly cares,
> Learn ye the potency of heaven-born mind,
> Its thought and life far from the herd withdrawn!
> Nearer the gods no mortal may approach.

No doubt the *Principia* has its faults. Given the short time-frame of its composition, it bears many human marks of haste. According to Whiteside, "the logical structure is slipshod, its level of verbal fluency none too high, its arguments unnecessarily diffuse and repetitive and its very content on occasion markedly irrelevant to its professed theme."[18] The German mathematician Leibniz, who attempted to reconcile the inverse-square law with Descartes's vortex theory, referred to Newton's gravity as "occult." How did it range across space without a medium to transmit its effects? For that matter, even Newton was dissatisfied with the rather mysterious concept of "action at a distance," and confided to Richard Bentley, master of Trinity, that it was "so great an absurdity, that I believe no man who has in philosophical matters a competent faculty of thinking, can ever fall into it."[19] He added, "the Cause of Gravity is what I do not pretend to know."[20]

Nevertheless, the superhuman scale of Newton's achievement remains. "Newton was engaged only about 18 months in the composition of his immortal work," observed Robert Grant. "When we contemplate, in connexion with this fact, the prodigious mass of original discoveries announced in the Principia, the mind is lost in amazement at the power of thought which could have reared into existence so stupendous a monument in such a brief space of time."[21] These discoveries included his demonstration of the inverse-square law of force for bodies moving in any of the conic sections (ellipse, parabola, and hyperbola); various theorems related to the attraction of the Sun, Moon, and planets; investigations of the shape of the Earth (oblate, i.e., flattened at the poles); and explanations of the tides and of the precession of the equinoxes. This last one was the grand phenomenon discovered by Hipparchus in antiquity, which had utterly baffled astronomers until Newton was able to show how it arose from the pull of the Moon on the equatorial bulge of the oblate Earth (an explanation so natural that, as George Biddell Airy later remarked, this part of the *Principia* "probably astonished and delighted and satisfied its readers more than any other").[22]

Of the problems causing Newton the greatest pains, one was the method of calculating the motions of comets. In the summer of 1686, he confided to Halley that he had not yet succeeded in solving the problem. He was baffled by the horrendous nature of the geometrical analysis. However, he devised a workable method by the time *Principia* went to the press.[23] Refinements were clearly needed, but Newton was able to calculate the parabolic orbit of the comet of 1680. Moreover, Newton's method was later adapted by Halley to calculate the orbits of a number of other comets, including the famous one that now bears his name.

The other problem with which Newton struggled heroically, and in the end unsuccessfully, was the difficult lunar theory. This is a special case of the theory that will be the underpinning of the present study, the theory of perturbations, the disturbing influence of one planet on the motion of another—the theory from which Vulcan later emerged as a seemingly inevitable consequence. In the case of the planets, "the then undeveloped state of the practical part of astronomy . . . had not yet attained the precision requisite to make such an attempt inviting, or indeed feasible."[24] Only in the case of the Moon was it otherwise. Here, after all, the disturber is the Sun itself, and the most important inequalities of the Moon's motion had already been obvious in the time of Hipparchus, 2,000 years earlier.

Basically, the Moon's orbit around the Earth is an ellipse. This is alternately compressed and relaxed by the disturbing effect of the Sun, thus forcing it to turn and wobble in various complicated ways. At times the Moon runs ahead, at other times behind the position it would have in an undisturbed elliptical orbit. The intricate motion of the Moon became the most critical test of the theory of universal gravitation—and Newton well knew it.

He conceived of the Moon's path as a Keplerian ellipse drawn from its ideal curve by the disturbing force of the Sun, a force always directed toward the Sun as it constantly changes its position relative to the Earth and the Moon. He could explain some of the

Moon's irregular motions at once, but he was baffled with one particular feature. The line between the Earth and the lunar perigee or apogee (the points where the moon is closest or farthest away) is known as the line of apsides. As was known long before Newton, this line advances in the same direction as the Moon's motion at an average rate of 3° 23′ per month, or 40° 41′ per year.

Qualitatively, Newton understood the cause of this advance very well. In the undisturbed ellipse that the Moon would follow if the Earth alone acted upon it, the orientation of the Moon's orbit would forever remain constant and fixed in space. The solar force destabilizes the situation, however. Averaged over the entire Moon's orbit, it decreases (by 1/357) the central pull of the Earth on the Moon. Thus the actual orbit of the Moon is slightly less curved than the undisturbed orbit, bringing the Moon to each successive perigee slightly later than it otherwise would. The line of apsides twists in the same direction the Moon is moving, and over successive orbits the Moon traces out an intricate rosette. Finally, after 8.9 years, the line of apsides returns to its original position. Unfortunately, when Newton attempted to carry out this analysis quantitatively, he was sharply disappointed; the value he found for the advance of the line of apsides came out to only 1° 31′ 14″ a month—just one half of the observed value. He had, as Westfall notes, "brought a prominent feature of the moon's motion hitherto recognized as an observed fact which defied explanation, within the scope of his celestial dynamics,"[25] only to be faced in the end with a glaring discrepancy. In the first edition of the *Principia,* he could not even bring himself to acknowledge his failure. The dynamical treatment of this important problem is noteworthy only by its singular omission.

In 1693, after the publication of the first edition of the *Principia,* Newton suffered a mental breakdown. His symptoms of paranoid delusions and depression may well have been partly induced by acute mercury poisoning (he had resumed his alchemical experiments, and often fell asleep by the still burning furnace). Whatever the cause, he resigned his Cambridge fellowship, which he had held for 30 years, and moved to London to become master and, later, warden of the mint. Despite this rather

abrupt change in the direction of his life, the lunar theory continued to obsess him.

In its grip, he turned, with increasing urgency and even des-
peration, to his colleague, John Flamsteed, who alone had the lunar observations that might allow him to achieve a breakthrough.
Flamsteed was born near Derby in 1646, the son of a prosperous
maltster. Crippled by an attack of rheumatic fever at 15, Flamsteed
remained physically delicate for the rest of his life, but never let
his handicaps interfere with his passion for astronomy, which was
awakened at an early age. After completing his degree at Cambridge, Flamsteed was appointed "His Majesty's Observator" at
the small observatory at Greenwich founded by Charles II in 1675
for the "finding out of the longitude of places for perfecting navigation and astronomy." At Greenwich, Flamsteed set out, with
great diligence and dedication, to rectify the places of the stars, especially those near the ecliptic "and in the Moon's way."[26] At first
he labored with rather inferior equipment, but he acquired, at his
own expense, a fine mural quadrant in 1689 to measure routinely
positions to an accuracy of only 10 arc seconds, compared to several minutes for those in Brahe's great catalog.

Beginning in summer 1694 Newton spent much of the year in
a final burst of concentrated effort on solving the theory of the
Moon. To assure success, he visited Flamsteed at Greenwich on
September 1, 1694, to request the necessary observational data.
The meeting was cordial; Flamsteed told Newton that he had a
plentiful store of observations, including "about 50 positions of
the Moon reduced to a synopsis," and promised a hundred more
should Newton ask for them.[27] Within a month Newton was asking. He was confident that with the right observations he might reduce the difference between theory and observation to only 2 or 3
arc minutes (Horrocks's lunar theory had been accurate only to
10). He hoped to complete it that winter. However, the observations he needed had not yet been made. Flamsteed, naturally
enough, was not eager to turn aside from his own work; to him,
Newton's requests for observations represented interruptions,
and were also increasingly curt and demanding. The two men
moved steadily and inevitably toward a rupture.

Even as late as April 1695, Newton was still optimistic of bringing the work to an end shortly, and wrote to Flamsteed that "the Moon's theory . . . I reccon . . . will prove a work of about three or four months." He added, as an indication of his mounting frustration, "when I have done it once I would have done with it forever."[28] The difficulties of the problem mounted, and as they did, Newton began to project more and more of his frustrations onto the Astronomer Royal. Flamsteed brought up a rumor he had heard that Newton was considering suppressing the lunar theory from the second edition of the *Principia* because of his refusal to supply observations. In his July 1695 reply, Newton mentioned several problems with which he had helped Flamsteed, then implied that the Astronomer Royal had failed to keep his side of the bargain: "I should never have undertaken [these things] but upon your account, & . . . I told you I undertook them that I might have something to return you for the Observations you then gave me hopes of, & yet when I had done saw no prospect of obteining them . . . I despaired of compassing ye Moons Theory."[29] This was decidedly unfair; Flamsteed had already sent Newton several hundred positions, and continued to provide still others. Flamsteed would later claim that he had spent all of 1695, when not ill, furnishing Newton with lunar positions. However, Newton was declaring moral victory by placing the blame for defeat squarely upon Flamsteed's shoulders.

It is a measure of Newton's defeat that in the end he could do no better than a model that Horrocks had produced: a rotating Keplerian ellipse of variable eccentricity. Newton smuggled it in as an expedient, adding some inequalities he had inferred from gravitational law, while others were taken from observation alone. The final result was disappointing. As Whiteside noted in his essay, "Newton's Lunar Theory: From High Hope to Disenchantment," his model was "only approximately justifiable from the 3-body dynamical problem whose more accurate solution he thereby ceased to control, then and ever afterwards."[30]

Newton was by now past 50 and must have known that his great creative period was over. His later years were absorbed with running the mint and presiding over the Royal Society. And yet he could not quite give up the obsession. Even as late as 1713, while revising the *Principia* for a second edition, Newton continued to make minor adjustments in his quasi-Horrocksian theory, but in vain. After Halley had finished a long series of observations of the Moon, Newton somewhat wistfully told John Conduitt "he would have t'other stroke at the Moon." In a soberer moment he realized it was too late. In another, equally revealing moment, when John Machin praised the "sagacity" of his published propositions on the lunar theory, Newton "smiled wanly and said his head never ached but with his studies on the Moon."[31] When Newton died in April 1727 at the age of 84, the problem remained as unsolved as ever.

three

Triumvirate

Interestingly enough, it was the French mathematicians, not the British, who adopted Newton's "Majestic Clockwork" as their own and brought it to a state of near-perfection, but not at first. For a half century after the *Principia*'s publication, Descartes's vortex theory (all space is filled with invisible particles of matter whirling in great eddies, or in French, *tourbillons*) reigned supreme. And yet, as Voltaire wrote, "the French always come late to things, but they do come at last."[1]

Voltaire himself played a decisive role in winning France over to Newtonian ideas. In 1736, he wrote a charming account of Newton's discoveries, *Éléments de la Philosophie de Newton,* which was an instant success. From 1740 onward, no more papers on the vortex theory ever appeared in transactions of the Paris Académie des Sciences.[2]

By then, French mathematicians were already well prepared to grapple with the complications of the theories of "le Grand Newton." Newton had derived many of the results of the *Principia* using the new branch of mathematics he had invented—fluxions. Essentially, he developed a method of investigating complex phenomena, such as planetary motions, by analyzing them in terms of infinitely small variations. In the *Principia,* however, he covered his tracks by recasting his derivations in standard geometrical form. Meanwhile, the German mathematician Gottfried Wilhelm Leibniz independently discovered some of the same mathematical methods, renamed them the infinitesimal calculus, and devised a more convenient system of notation. Later innovators, such as the Swiss Jacques Bernoulli and Leonard Euler, developed the Leibnizian calculus into a modern and far-reaching instrument and refashioned the difficult Newtonian demonstrations into more straightforward operations involving differential equations and algebra.

The most pressing problem, and still the Everest of Newtonian dynamics, was the difficult lunar theory. Newton's Horrocksian "corrections" of 1694–1695 proved a dead end. It was clear to his successors that there was no alternative but to start over from first principles. Three mathematicians, the Swiss Euler and the French Alexis-Claude Clairaut and Jean le Rond d'Alembert, took up the challenge.

All three succeeded in reducing the three-body problem of the Moon's motion to three differential equations. However, they were unable to solve them directly. Clairaut wrote in despair of his equations that anyone "may now integrate them who can. . . . I have deduced the equations given here at the first moment, but I only applied a few efforts to their solution, since they appeared to me little tractable. Perhaps they are more promising to others."[3] The others, however, found them no more tractable. They remained unsolved, and there still is no exact solution.

Under the circumstances, the only choice was to attempt to approach the solution by means of successive approximations. The method involves beginning with the unperturbed orbit of the perturbed body, the Moon or a planet, and then calculating the forces and accelerations due to the perturbing body (in the case of the Moon, the Sun). This produces a slight change in the position of the perturbed body, leading to a further small increment in the forces and accelerations (second-order deviations, so-called because they are small compared to their first values), and so on. In principle, one can carry the approximations as far as one wishes.

However, "since in the higher orders all these terms react upon one another, their complete computation constituted from the beginning an entangled, nearly inextricable task, demanding years of work—in later times with higher standards, even an entire life of strenuous and careful work."[4]

Approached using such methods, even the solution of the lunar apsides problem seemed within reach. Each investigator used a somewhat different method, but each arrived at the same disappointing result: The line of the apsides appeared to precess at half the observed value. Theory and observation remained in disagreement. A moment of truth had arrived. Universal gravitation was faced with an acute crisis. In September 1747, Euler notified Clairaut: "I am able to give several proofs that the forces which act on the moon do not exactly follow the rule of Newton."[5] With this encouragement from Euler, Clairaut, too, became convinced of the need to abandon the exact inverse-square law, and proposed replacing it with an inverse-square plus a higher-order term. Only d'Alembert hesitated: "I am so much afraid of making assertions on such an important matter that I am in no hurry to publish," he wrote. "Besides I will be very sorry to overthrow Newton."[6]

Two years later, Clairaut nerved himself to carry out a more rigorous computation, taking account of some small terms he had hitherto ignored. One had to do with the motion of the Sun. Since the Moon's motion relative to the Earth is so much more rapid than the Sun's, the position of the latter can, to a first approximation, be regarded as fixed. This had been the approximation Newton had used. However, it turned out that the effects of the Sun's motion are not so negligible as might be thought. A detailed analysis showed the overall advance of the line of the apsides is produced by the excess of the disturbing effect of the Sun on the Moon at apogee over that at perigee. Near apogee, the line of apsides is advancing. Since the Sun's motion is in the same direction, the Sun tends to keep up with it, increasing its effect. At perigee, the line is regressing, but the Sun's motion is in the opposite direction, decreasing its effect. Perturbation augments perturbation, and when these second-order effects are summed up, they are found to contribute just enough to bring the calculated advance of the line of the apsides into exact agreement with the observed value.[7] There was, then, no longer any reason to introduce a correction to the Newtonian theory. Clairaut had found the grail. "Thus," notes Grant, "a circumstance, which at one time threatened to subvert the whole structure of the Newtonian theory, resulted in becoming one of its strongest confirmations."[8]

Flushed with this triumph Clairaut could well have rested on his laurels. But in 1757, he successfully completed a computation of the lunar and planetary perturbations of the Earth. At once, he began to stir to an even grander challenge: the refinement of the date of the forthcoming return to perihelion of the comet that Edmond Halley, dead since 1742, had prophesied in 1758-1759. Clairaut resolved to snare it within the web of his new perturbation theory.

Halley had realized the comet, whose previous returns to perihelion had occurred in 1531, 1607, and 1682, closely approached Jupiter in 1681. The comet, as a result of its gravitational acceleration, was forced into a somewhat larger orbit; thus its next return would be delayed, but by how much? Halley's best guess was "it is probable that [the comet's] return will not be until . . . about the end of the year 1758, or the beginning of the next." He added, "Wherefore if according to what we have already said it should return again about the year 1758, candid posterity will not refuse to acknowledge that this was first discovered by an Englishman."[9]

Clairaut's method was brutally straightforward. He believed one needed to calculate successive positions of the comet in its orbit at intervals of several days, the distance separating the comet at each point from the massive planets Jupiter and Saturn, and the resulting effects on the comet's motion and period. Given the tight timeline—the comet, after all, was approaching steadily and ever more rapidly to perihelion—there was no time to lose. Clairaut would never have attempted the overwhelming mass of calculations alone, and so engaged two assistants, the young astronomer Joseph-Jérôme Lefrançais de Lalande and the gifted and courageous Nicole-Reine Étable de la Brière Lepaute, a clockmaker's wife.[10]

Problems of gravitational analysis showed a remarkable propensity to become more intricate than they appeared at first. So it had been with the lunar theory, which Newton had once hoped to master in a matter of months. On closer acquaintance, the comet problem also required more effort. Initially Clairaut assumed it would be unnecessary to take account of any perturbing effects on the comet except during its close passages by Jupiter and Saturn. He was dismayed to find out that for the calculation to be reliable, their effects would have to be summed over an entire orbit. Moreover, even Jupiter's pull on the Sun could not be ignored. Since a small displacement of the Sun's position could have a large effect on the size of the comet's orbit, hence of its period of revolution, the calculations would have to be carried out over two full orbits of the comet—a period of 150 years. This covered the period back to the comet's two previous returns: 1607 and 1682. To further check the technique's accuracy, Clairaut decided to extend the calculation back over one more orbit, to its return in 1531. The positions of the comet and Jupiter and Saturn, together with the effects of the pertubations of these planets on the comet's orbit, had to be calculated over 700 times. In all, the triumvirate worked diligently for 6 months, usually from early morning until late at night, sometimes not even pausing from their calculations during meals. Lalande later claimed that as a result of the strain of these extraordinary efforts, he contracted an illness that changed his temperament for the rest of his life.[11]

By November 14, 1758, Clairaut reached a tentative result and presented it to the Paris Académie des Sciences. The immense calculations had been worked through completely for Jupiter but not yet for Saturn; the latter could only be estimated. They indicated the comet would reach perihelion within a month either side of mid-April 1759. Their uncertainty was based on the error in the calculation of the perihelion passage of 1682.

Even then it was being searched with a 4 ½-foot Newtonian reflector from the observatory on the roof of the Hotel Clûny in Paris, by the noted scourer of the skies, Charles Messier, who was employed under Joseph Nicolas Delisle. He knew of Halley's prediction, but there is no evidence he took any account of Clairaut's revision. During much of the winter of 1758 the skies were cloudy over Paris, but on the clear evening of January 21, 1759, he picked up a suspicious object in Pisces. With some difficulty, Messier and Delisle contained their excitement, wanting to be certain that it was indeed the comet, not merely an unexpected interloper. They followed it until mid-February, when it disappeared in the evening twilight, not to reappear into the morning sky until April.

Up to this point, Messier and Delisle apparently thought they had the comet all to themselves. Given the efforts of their countrymen to calculate the comet's return, it would have been fitting had its actual discovery fallen to a Frenchman. They later learned a German farmer, Johann Georg Palitzsch, who lived at Prohlis, near Dresden, had already picked it up on Christmas night 1758.

Throughout April and May 1759, Halley's Comet, as it came to be known henceforth, was widely observed. Orbits calculated by Lalande and others showed that it had passed perihelion on March 13—thus the prediction of Clairaut and his collaborators was off by only 33 days. Once again, universal gravitation had proved resoundingly triumphant. "The return of the Comet of 1682 in the time prescribed by the Newtonian theory is one of the brilliant triumphs of physics and marks an epoch in this science," Clairaut himself exulted.[12] The "lawless" comets had been subordinated to law. The promise that Halley had poetically captured in his tribute to Newton affixed to the *Principia* had been abundantly fulfilled:

> Now we know
> The sharply veering ways of comets, once
> A source of dread, nor longer do we quail
> Beneath appearances of bearded stars.

After his famous Halley's Comet prediction, Clairaut was in such demand in the fashionable salons that it interfered with his scientific work. "Engaged," says Bossut, "with suppers, late nights, and attractive women, desiring to combine pleasures with his ordinary work, he was deprived of his rest, his health, and finally, at the age of only 52, of his life."

Meanwhile his rival d'Alembert had given up mathematics for letters. With Diderot and Voltaire, he became the driving force behind the great *L'Encyclopédie*. It is for this, rather than for any mathematical equation, he is best remembered today. Carlyle said of him that he was "an independent, patient, and prudent man; of great faculty, especially of great clearness and method; famous in Mathematics; no less so, to the wonder of some, in the intellectual provinces of Literature."[13] When d'Alembert died in 1783, Diderot, who was at that moment also on his deathbed, remarked, "A great light has gone out."

four

The Celestial Decades

The torch now passed to a new generation of mathematicians, of whom the greatest were Joseph Louis Lagrange and Pierre Simon de Laplace. Lagrange was born in Turin in 1736. His father was very well to do and at one time controlled the Sardinian war chest. He eventually lost most of his property by speculation, so that by the time young Lagrange passed to manhood there was little of his fortune left. "If I had inherited a fortune," Lagrange later noted, "I should probably not have cast my lot with mathematics."[1] He wrote his first important paper when he was only 19, and by the time he reached 25 had become the greatest living mathematician. "In appearance he was of medium height, and slightly formed, with pale blue eyes and a colorless complexion. In character he was nervous and timid, he detested controversy, and to avoid it willingly allowed others to take the credit for what he had himself done."[2]

In 1749, his great rival, Laplace, was born at Beaumont-en-Auge, in Lower Normandy, to peasants. Little is known of his early years, since after becoming well-known he did everything he could to distance himself from his relatives and those who had assisted him. At 18 he was sent to Paris with recommendations from several prominent people. On arriving in the capital, he called on d'Alembert, sending his recommendations ahead of him; d'Alembert refused to see him, saying that he was not interested in young men who came only with recommendations by prominent people. Laplace returned to his lodgings and wrote d'Alembert an essay on the principles of mechanics, whereupon d'Alembert, completely disarmed, invited him back, saying, "Sir, you see that I paid little enough attention to your recommendations; you don't need any. You have introduced yourself better."[3] Through d'Alembert's influence, Laplace obtained a position as professor of mathematics at the École Militaire of Paris.

Clairaut's solution of the problem of the lunar apsides and the refined prediction of the return of Halley's comet inspired renewed confidence in the ability of Newton's theory to satisfy all the phenomena of the planetary motions. Among the outstanding remaining problems was the "great inequality of Jupiter and Saturn." Long before, Kepler and Horrocks had recognized there was something seriously amiss in the motions of these planets. Flamsteed and Halley confirmed their suspicions: The mean motion of Jupiter appeared too slow, that of Saturn too quick. To correct the problem, Halley, in his 1695 tables, added a regular acceleration of Jupiter of $0° 57'$ per thousand years, and a retardation of Saturn of $2° 19'$. The problem of explaining the great inequality was proposed as the subject for the prize of the Paris Académie des Sciences in 1748 and 1752, both won by Euler, who despite taking novel approaches failed to solve the problem.

These unsuccessful attempts nevertheless served their purpose by directing the attention of astronomers to the so-called secular inequalities of the planets—slow variations in their elliptic elements due to perturbations of the other planets taking place over very long periods of time. Since the secular variations proceeded always in the same direction, and did so to apparently indefinite extent, they must in time destabilize the solar system.[4] Newton, indeed, had surmised that this was so. He believed divine intervention might be needed from time to time to restore it to order.

The problem of the great inequality of Jupiter and Saturn was solved in 1784 by Laplace. Five times the mean motion of Jupiter is a little more than twice that of Saturn: The periods are nearly commensurable. This means Jupiter and Saturn line up at nearly the same longitude in their orbits after 5 Jovian and 2 Saturnian years, or once in every 59 terrestrial years. The cumulative effect of this periodic disturbance is, as Laplace surmised, a very large perturbation.[5] Since the periods are not quite exactly commensurable, each conjunction takes place at a longitude a little greater than the preceding one, with the point of conjunction making a complete orbit in about 900 years. This corresponds to an average inequality in the case of Saturn equal to 48′ 44″ per century and in that of Jupiter of 20′ 49″ per century in the opposite direction. The apparent retardation of Saturn and apparent acceleration of Jupiter reached a maximum in 1560, then each began to decrease toward the value of the mean motion, which they attained in 1790. Thereafter the reciprocal effect of the two planets on one another reversed direction, with the motion of Saturn accelerated and that of Jupiter retarded.

Of his achievement in explaining this intricate planetary dance, Laplace later wrote:

> When I had recognized these various inequalities, and examined more carefully than had been done before, those which had been submitted to calculation, I found that all the observed phenomena of the motions of these two planets adapted themselves naturally to the theory; before they seemed to form an exception to the law of universal gravitation; they are now become one of the most striking examples of its truth. Such has been the fate of this brilliant discovery of Newton, that every difficulty which has arisen, has only furnished a new subject of triumph for it.[6]

Two more years of calculation led Laplace to the result that, whatever the relative masses of the planets, the eccentricities and inclinations of the planetary orbits would always remain nearly the same, provided only that the planets all revolved around the Sun in the same direction. Lagrange had already discovered that the mean distances (semimajor axes) of the planets are essentially constant, being subject only to slight periodic changes.[7] Solar system deviations were, then, apparently confined within narrow limits. Like a great clockwork, it would glide on, in accordance with the law of gravitation, a vast self-correcting machine, eternal and predictable. In reality, this result was illusory; the theories proposed by Laplace and Lagrange were less assured than they thought, but they exerted considerable influence, the uncertainties of such intricate calculations remaining to be fathomed.[8] Their limitations were as yet insensible to Robert Grant, who wrote in his famous 19th-century *History of Physical Astronomy:*

> These laws which thus regulate the eccentricities and inclinations of the planetary orbits, combined with the invariability of the mean distances, secure the permanence of the solar system throughout an indefinite lapse of ages, and offer to us an impressive indication of the Supreme Intelligence which presides over nature, and perpetuates her beneficent arrangments. When contemplated merely as speculative truths, they are unquestionably the most important which the transcendental analysis has disclosed to the researches of the geometer.[9]

Perhaps all the phenomena of nature could eventually be subsumed under such deterministic laws, as Laplace believed and summed up in his famous boast: Given the initial position and motion of every particle in the universe, and an intelligence vast enough to comprehend them, "nothing would be uncertain. Both future and past would be present before its eyes." For him, all phenomena were reducible to a problem of mechanics. To some this was a sublime vision, to others quite the opposite, since it seemed to indicate a system closed and oppressive. So it became for Thomas Carlyle who, while noting that "our Theory of Gravitation is as good as perfect," added, admonishingly: "System of Nature! To the wisest man, wide as is his vision, Nature remains of quite infinite depth, of quite infinite expansion; and all Experience thereof limits itself to some few computed centuries and measured square-miles. . . . We speak of the volume of Nature. . . . It is a Volume written in celestial hieroglyphs . . . . [and yet] that Nature is more than some boundless Volume of . . . Recipes, or huge, well-nigh inexhaustible Domestic-Cookery Book, . . . the fewest dream."[10]

Among the celestial recipe-makers was Laplace. Delambre was another. He attempted to produce improved tables of plane-

tary motion from Laplace's formulae. Delambre's moment of revelation came in 1786, the year Lalande had predicted a transit of
Mercury. On May 3, the date assigned, the Sun's disk was watched
carefully by all the savants of Paris. Delambre wrote:

> At sunrise it rained; all the astronomers of Paris were at their tele-
> scopes, but, fatigued with waiting, and no longer retaining any
> hope, they quitted their places half an hour after the time an-
> nounced for the planet's egress from the Sun's disk. I resolved to
> wait till the moment indicated by Halley's tables; but such a degree
> of perseverance was unnecessary, for the phenomenon took place
> three quarters of an hour later than the time fixed for it by Lalande,
> and three quarters of an hour earlier than that assigned by the ta-
> bles of the English astronomer.[11]

Lalande was so annoyed by this error in his prediction that he
at once resumed his researches on the planet, while Delambre resolved to take full account of the perturbations arising from the
mutual attractions of the planets in preparing new tables of their
motions. Within the week Delambre offered his services to
Laplace to compute new tables of Jupiter and Saturn, which
would incorporate the just-announced theory of the great inequality. At the time, no one doubted the Newtonian theory
would prove adequate to the task. As Grant observed, the experience had been that "these anomalies, one by one, have yielded to
the researches of the geometer, and, the law of gravitation still re-
tains the character of simple grandeur by which it was distinguished, when first announced by its immortal discoverer."[12]

In part Delambre and his successors were successful, so much
so that in 1796, Laplace, in his *Exposition du système du monde*,
could announce with the complacency of someone looking back
with satisfaction on a completed project:

> It is only three centuries since Copernicus first introduced into the as-
> tronomical tables the motion of the planets around the Sun . . . [and
> yet] these tables have acquired a degree of precision which could
> never have been anticipated; formerly their errors amounted to sev-
> eral minutes, they are now reduced to a small number of seconds,
> and very often, it is probable, that their apparent deviations arise
> from the inevitable errors of the observations.[13]

Against this sanguine picture of steady and assured progress, the discrepancies in Mercury's motion would remain unresolved—despite all of Lalande's efforts to reduce them to order. For the present, this was no more than a faint tarnish on the fair fame of Newtonian theory.

Meanwhile the frigid planet beyond Saturn that William Herschel had unexpectedly discovered in 1781 had also begun to experience unexplained divagations. In a few years, these cracks would appear as potential fracture lines along which the whole Newtonian edifice threatened to crumble. It was to the attempt to repair these cracks that the greatest of Laplace's successors, Le Verrier, would devote the efforts of a great career and come to occupy a privileged position in the unfolding drama of planetary discovery.

PLANETS DROWNED IN NIGHT

five

Nebulous Star or Comet?

As astronomers worked out the orbits of the planets, it scarcely occurred to them that the system they had attempted to reduce to calculation might be incomplete. Occasionally the idea surfaced, only to be rejected. Kepler had made the suggestion there might be additional planets, one between Mercury and Venus, a second between Mars and Jupiter—a vision that, in the latter case at least, would one day come to seem prophetic. But after being captivated by the hypothesis of the regular solids that he described in his early work *Mysterium Cosmographicum*, even he had no more reason to believe there were more than five planets.

In 1609, the same year Kepler announced the discovery of the elliptic motions of the planets in his *Astronomia Nova*, Thomas Harriot and Galileo pointed the first telescopes at the heavens. The initial object of their attention was, naturally enough, the Moon. Then early in 1610, Galileo discovered the four "Medicean Stars" revolving around Jupiter.

In his hastily written account, Galileo described these hitherto unglimpsed orbs as "small planets," and named them for his patron Cosimo de'Medici II. The creation had not yet been fully inventoried, evidently; there were other worlds, perhaps, for all anyone knew to the contrary, inhabited by other Galileos. And yet for all their revolutionary implications, the "Medicean Stars" occupied a clearly subordinate position in the scheme of things; they revolved around Jupiter as the Moon did around the Earth, so that instead of being true planets, as Galileo inferred in the first flush of enthusiasm, they were really moons or "satellites," the term that Kepler coined for them. All the same their existence vouchsafed that besides the Sun there were at least two other centers in the universe: the Earth with its Moon and Jupiter with its more impressive retinue of satellites. The Earth-centered fantasy of Ptolemy could no longer be decently maintained (although it continued to enjoy dogmatic status within the Church).

Other observers built better telescopes and discovered more moons. Christiaan Huygens of Holland added Saturn's largest moon, now known as Titan, in 1655, and his sharp-eyed rival Giovanni Domenico Cassini at the Paris Observatory discovered four more Saturnian satellites between 1671 and 1684 (to say nothing of the nonexistent moon of Venus, which Cassini claimed in 1686; it continued to haunt Venus observers for more than a century).[1]

Before the establishment of the existence of sunspots as actual blemishes on the solar surface, there were reports of intramercurial worlds as well, perhaps fleets of them, as recorded in the "Bourbonian Stars" of Canon Tarde. Long forgotten, they would enjoy a brief revival during the heyday of the Vulcan controversy. That something might exist at the other extreme, beyond the orb of Saturn, the traditional realm of the vault of fixed stars, seems to have occurred to no one. For all the advances brought by gravitation and the telescope to the understanding of the motion and physical nature of the bodies of the solar system, the medieval cosmology remained firmly entrenched in the human imagination of the sidereal universe.

That the space between the orb of Saturn and the fixed stars should be void was a conclusion that followed no logical necessity; to the contrary, it was apparent there were denizens that wandered through its mysterious regions. Halley had shown the comet named for him followed a highly elongated orbit that took it well beyond the orbit of Saturn. While in more careful tracking across the orbits of Jupiter and Saturn and thence into the unexplored reaches of outer space Clairaut had glimpsed, with prescience, that the wayfarer might be further perturbed by worlds unknown. "A body which travels into regions so remote," he declared to the Paris Académie des Sciences in November 1758, "and is invisible for such long periods, might be subject to totally unknown forces, such as the action of other comets, or even of some planet too far distant from the Sun ever to be perceived."[2]

Despite such prophetic utterances, no one was prepared for what would be revealed to a little-known amateur astronomer on the fateful night of March 13, 1781. Friedrich William Herschel (Figure 6), musician of the city of Bath, England, had already completed one telescopic "review" of the heavens and was diligently employed that night, as he had been since August 1779, on a second examination.

Herschel was born in Hanover in 1738. His father, Isaak, was an oboist in the Hanoverian Guard, and young Herschel followed in his footsteps. But the military life did not appeal to him. After coming under fire at the battle of Hastenbeck in 1757, he left Hanover for England (where he later naturalized his name to William).

Herschel came at an opportune moment. It was the era of the personal union of Hanover with Great Britain that had begun with the reign of George I in 1714 (and would continue until Victoria's reign in 1837). George II was on the throne, and there were unprecedented opportunities for talented Hanoverians. For several years Herschel led a somewhat unsettled existence, finding employment as a copyist, teacher, performer, and composer of music. They were years of struggle, but he impressed wherever he went, and finally obtained an important post—organist at the Octagon Chapel in the fashionable resort town of Bath. He now had a good and reliable income, which he supplemented by giving music lessons and concerts. His energy was inexhaustible, for at this time he also became interested in astronomy. Finding that commercially available telescopes were too expensive, he decided to build his

6. William Herschel (1738–1822). (Courtesy of Yerkes Observatory.)

own of the reflecting type, that is, using a mirror instead of a lens to gather light. At first he had many failures, but he finally succeeded in finishing a telescope with a good 4 ½-inch mirror, which he used to make his first review of the heavens. This was supplanted by a 6 ½-inch reflector that in quality surpassed the instruments then in use at the Royal Observatory at Greenwich.

It was the larger telescope he was using on March 13, 1781. He was in quest of the distance to the stars. A few stars lying next to each other had been discovered by other astronomers before Herschel had begun his more thorough and systematic research. The suggestion had even been put forward that the stars of a pair might be related like a planet and its satellite. More commonly they were supposed to be "optical" doubles—stars only accidentally juxtaposed along a given line of sight. If the latter, one star of a pair must in general be farther away than the other, perhaps much farther away. If observed from opposite sides of the Earth's orbit, one of the stars might betray a shift in its position relative to the other—a *parallax* from which its distance might be determined. At the moment, this grand object was what Herschel had in his sights: to capture as many double stars as possible and use them for nothing less than to take soundings of the immensity of space.

Between 10 and 11 o'clock, he came to the interesting region between the Crab Nebula in Taurus and the open cluster Messier 35 in Gemini. There was something peculiar about one of the stars in the field. It appeared slightly larger than its neighbors and had an unusual luster. When inspected with high magnification, it appeared "hazy and ill-defined."[3] "In the quartile near ζ Tauri," Herschel jotted down in his observing book, "the lower of two is a curious either nebulous star or perhaps a comet." Noting this peculiarity, he passed on, and without skipping a beat returned to his search for double stars, which he continued until 5 o'clock the following morning.

Herschel suspected he had discovered a comet. This was interesting, no doubt, but hardly momentous. Not until Saturday, March 17 did he have an opportunity to reexamine the field. What he found only confirmed his suspicions. Looking at his comet or nebulous star, he declared emphatically, "it is a comet, for it has changed its place." Using his micrometer, he obtained a first measure of the diameter of its tiny disk. The following night he showed the "comet" to his friend William Watson, and took another measure. Herschel announced his discovery in a letter to the Royal Society. He kept up the observations well into April, never doubting the object's cometary nature, despite the fact its disk always appeared perfectly defined, "having neither beard nor tail." His measures disclosed, moreover, an apparent increase in the diameter of its disk, "from which we may conclude," he noted in his log, "that the Comet approaches us."[4]

But Nevil Maskelyne, Astronomer Royal, was puzzled: "it is a comet or a new planet, very different from any comet I ever read any description of or saw." As Maskelyne worked out a provisional orbit for the object, he found, surprisingly, what Herschel had found was "as likely to be a regular planet moving in an orbit very circular around the Sun as a comet moving in a very eccentric ellipsis."[5] But the idea was too unsettling; he, too, changed his mind, influenced in part by Herschel's series of disk measures, and opined that "it is coming down to the Sun and that it probably will acquire a surrounding coma and tail," thereby declaring definitively its cometary nature. On April 29, the ferret of comets himself, Charles Messier, wrote to congratulate "Monsieur Hertsthel at Bath" on his discovery, adding that "nothing could be more difficult than to recognize the new comet, and I cannot conceive how you were able to return several times to this star—or comet . . . since it had none of the characteristics of a comet."[6]

Hereafter Herschel only too gladly yielded the continued study of his problematic object to the professionals, and the mathematicians set to work trying to calculate precisely the orbit. Starting from the assumption that the object was a comet, moving in a parabolic orbit, failure was preordained.

Pierre François André Mechain, Messier's chief rival, announced a provisional orbit that had the "comet" reaching perihelion, at a distance from the Sun of only half that of the Earth, by the end of May. Something was seriously wrong. Another French mathematician, Jean Baptiste Gaspard Bouchart de Saron, startled the Académie des Sciences on May 8, 1781, by announcing the object was actually more distant than previously supposed—at least as far out as 14 astronomical units (AU). If so, it was orbiting far beyond Saturn, and was therefore in all probability a planet. Abandoning altogether the attempt to reduce the object's motion to a parabolic orbit, Anders Johann Lexell, professor of mathematics at St. Petersburg, utilized two observations—one by Herschel made on March 17, the other by Maskelyne on May 11—to successfully fit a circular orbit. Lexell's computations showed the object had a mean distance from the Sun of 19 AU and a period of something more than 82 years. This settled the matter. As Lalande was later to remark, "from that time, it appeared to me that the body ought to be called the new planet."[7] In some sense, Saron and Lexell are entitled to be called the discoverers of Uranus; they were the first to break clearly with tradition in declaring decisively that Herschel had discovered a world beyond the orbit of Saturn.

The significance of this recognition dawned only gradually on the astronomy world. While the discovery of a new comet was something to be expected in the natural course of things, a new planet was another matter. None had ever been recognized in all recorded history; indeed, its existence broke completely with the formulations of orthodox thought, and it took time to adjust to the new outlook.

Lalande was among the first to embrace the conclusion and suggested the planet be called "Herschel." Johann Elert Bode of Germany was another. He proposed the name "Uranus."[8] Others attempted to balance excitement with caution. Herschel's friend William Watson, a fellow member of the Bath Philosophical Society, wrote to the discoverer in late August: "Your success, you perceive, has set both the English and French astronomers on exerting themselves to the utmost; a *comet* which is likely to be visible for at least 12 years is a very extraordinary phenomenon."[9] Herschel himself was singularly shy of the larger implications. In November, the president of the Royal Society, Sir Joseph Banks, announced that the society planned to award Herschel its Copley medal, and added: "Some of our astronomers here incline to the opinion that it is a planet and not a comet." Herschel, however, insisted: "With regard to the New Star I may still observe that tho' we are not sufficiently acquainted with it to ascertain its nature, yet enough has been seen already to shew that it differs in many essential particulars from Comets and rather resembles the condition of Planets."[10]

Banks presented Herschel with the Copley Medal in late November, and at last prophetically outlined the grander prospects: "Who can say but what your new star, which exceeds Saturn in its distance from the sun, may exceed him as much in magnificence of attendance? Who can say what new rings, new satellites, or what other nameless and numberless phenomena remain behind, waiting to reward future industry?"[11]

Herschel's discovery of the seventh planet brought him fame, and eventually royal patronage. Though a first attempt shortly after the discovery to obtain for him a place at King George III's observatory at Kew came to naught, largely because it had already been promised to another, Herschel and his friends contrived to keep his name before the king. In May 1782 the intrigue became more intense. Hitherto Herschel, despite the urging of Banks and others that he assert his prerogative as discoverer to name the new planet, had shown little interest in that question. Bode's name of "Uranus" was now coming into general use among European astronomers, but Herschel, who until now had referred to the object simply as "my planet," belatedly entered the arena and proposed naming it after George III. In early July he went to Windsor to show the king and queen Jupiter, Saturn, and several other objects through his telescopes. The king was duly flattered by this attention and decided to grant Herschel a pension of £200 per year, enabling him to give up music and devote all his time and energies to astronomy. His duties as "royal astronomer" (not Astronomer Royal; the position at the time occupied by Nevil Maskelyne) required him to live closer to Windsor. For this purpose he and his sister Caroline, who assisted him in his work, moved from Bath to Datchet, in the Thames Valley, where he set up his observatory. It was here that he began his "third review" of the heavens, discovering that the double stars he had been so carefully cataloging in the hopes of using them to derive the distances to the stars were, in fact, gravitationally bound to one another—a proof that universal gravitation extended far beyond the solar system. He also expanded his general study of the heavens by carefully cataloging the clusters and nebulae of the deep sky. It is said his landlady's ill temper forced him to move soon afterward from Datchet to

Slough, where he continued his reviews of the heavens. There he set up his largest telescope, the celebrated 40-foot reflector that had a 48-inch speculum.

Soon after receiving the pension from George III, Herschel officially announced to Banks his proposal of the name *Georgium Sidus*, the Star of the Georges, for "the new star which I had the honor of pointing out" and was universally recognized as "a Primary Planet of our Solar System":

> . . . as a subject of the best of Kings, who is the liberal protector of every art and science;—as a native of the country [Hanover] from which this illustrious Family was called to the British throne;—as a member of that Society which flourishes by the distinguished liberality of its Royal Patron;—and last of all, as a person now more immediately under the protection of this excellent monarch, and owing everything to his unlimited bounty; I cannot but wish to take this opportunity of expressing my sense of gratitude, by giving the name Georgium Sidus . . . to a star, which (with respect to us) first began to shine under His Auspicious Reign.[12]

Posterity cannot, perhaps, share Herschel's assessment of George III, who was hardly the most successful of the English monarchs. Winston Churchill characterized him as typically Hanoverian, capable of mastering detail, but limited at dealing with large issues and main principles.[13] Undoubtedly he is best remembered for presiding over the loss of the American colonies in 1781 and for the bouts of madness that occurred at intervals during his reign. After 1811, he was permanently deranged; one of his delusions was of seeing Hanover from Windsor Castle through Herschel's telescope. And yet he could find consolation in the observation of Matthew Turner that "it is true we had lost the *terra firma* of the Thirteen Colonies in America, but we ought to be satisfied with having gained in return by the generalship of Dr. Herschel a *terra incognita* of much greater extent *in nubibus*."[14]

The name Georgium Sidus did not survive. It continued to be used by Herschel himself. In Great Britain, it was still officially recognized by the *Nautical Almanac* until the middle of the next century. On the continent, Bode's mythology-based name "Uranus" prevailed.

In one sense, Herschel's discovery of Uranus was accidental. He was not searching for a new planet, and indeed only gradually recognized the significance of the object that he had found. In another sense, however, the discovery was a direct result of his unique plan of attack, of his attempt to compile a "natural history of the heavens." His project was, after all, to examine with an informed eye the contents of each field of the eyepiece, to carefully mark down anything unusual or out of the way. No one else had ever attempted anything like this except Messier, who had been keeping a list of nebulous objects found during his comet searches, not because of any interest inherent in these objects, but because they were apt to be mistaken for the comets that he sought with such single-minded intensity.

In September 1782 Herschel remarked to Lalande that "the discovery was not owing to chance" since the planet "must sooner or later fall in to my way, and as it was that day the turn of the stars in that neighbourhood to be examined I could not very well overlook it."[15] This was a point he was at considerable pains to emphasize, and did so repeatedly, later telling Charles Hutton:

> It has generally been supposed that it was a lucky accident that brought this star to my view; this is an evident mistake. In the regular manner I examined every star of the heavens, not only of that magnitude but many far inferior, it was that night its turn to be discovered. I had gradually perused the great volume of the Author of Nature and was now come to the page which contained a seventh planet. Had business prevented me that evening, I must have found it the next, and the goodness of my telescope was such that I perceived its visible planetary disc as soon as I looked at it; and by the application of my micrometer, determined its motion in a few hours.[16]

And yet this account is itself, as Simon Schaffer calls it, "an understandable exaggeration."[17] It gives short shrift to the actual sequence of events by glossing over Herschel's long indecision, his delay for the better part of a year before he accepted Uranus's planetary nature. Nevertheless, the essential point is valid, since with Herschel's refined telescopic optics and keen eye the discrepant appearance of the "star" came immediately and inescapably to his notice. Others had passed it by, not once but many times, without having the least suspicion aroused. Indeed, before Herschel had finally and firmly seized it, Uranus had been passed over as an ordinary star on no fewer than 23 occasions.

The first "prediscovery" observation was made December 23, 1690, when John Flamsteed, diligently mapping stars in the "way of the moon," unsuspectingly entered it in his catalog as a star, 34 Tauri. Flamsteed made six more observations of it between 1712 and 1715, including three in the course of a single week.[18] James Bradley, the third Astronomer Royal (after Flamsteed and Halley) recorded it as a star in 1753.[19] So did the German astronomer Tobias Mayer, in 1756. However, a French observer, Pierre Charles Lemonnier, holds the record.

Lemonnier was a distinguished astronomer and professor of physics in Paris, who in 1736 had traveled with Maupertuis and Clairaut to Lapland to measure an arc of the meridian—the expedition that verified in the most striking way Newton's calculation of the oblate shape of the Earth. Like many astronomers of his day, he was interested in improving the tables of the Moon, and with this purpose set out to map the stars of the zodiac with the mural quadrant of the observatory of the Capuchines in Paris. While doing so, he recorded Uranus on no less than 12 occasions between 1750 and 1771, including six times in a 9-day period in 1769. Four of his observations were made on consecutive nights (January 20–23), but the planet was then passing near its stationary point, and its meager daily motion went unnoticed. Indeed, it was practically unnoticeable. And yet even to think of comparing his observations was out of Lemonnier's way. He was intent on star-mapping. His instrument, though suitable for positional astronomy, was not powerful enough to show the planet's tiny disk. So self-absorbed, Lemonnier passed it by, unconscious of the significance of what he had measured. Only in 1788–1789, long after Herschel's discovery, did he return to his observations and with a heavy heart realize what a pearl of great price had sifted through his hands.[20]

Lexell's circular orbit was, of course, only a preliminary calculation, and there was every reason to expect that like the other planets, Uranus would follow an elliptic path, and moreover, that this path would be perturbed by the other planets. The earliest attempts to work out a refined orbit were made by Barnaba Oriani of the Brera Observatory of Milan and by the great Leonhard Euler (who also happened to be Lemonnier's father-in-law), who outlined his calculations at dinner with Lexell and his family just before he died on September 18, 1783.

Certainly, for all the unexpectedness of its arrival, like a foundling on the rear porch of the solar system, Uranus seemed to fit nicely into the general scheme of things. For one thing, it was at just the distance expected according to a strange, almost fanciful, relationship known as "Bode's law." Bode in his 1772 primer *Deutliche Anleitung zur Kenntniss des gestirnten Himmels* (Clear Introduction to the Study of the Starry Sky) had brought to general notice a curious fact, a fact that had actually been noted a few years earlier by Johann Daniel Titius, professor of mathematics at Wittenberg.[21] What Titius had found was that if one takes the numerical series 0, 3, 6, 12, 24, 48, and 96, and adds 4 to each term, the resulting set of numbers 4, 7, 10, 16, 28, 52, and 100 closely approximates the relative distances of the planets from the Sun. There was no a priori reason why the scheme ought to work, and yet work it did and remarkably well, with one glaring exception. In the position of number 28, there appeared to be a gap. These results are summarized in the table below:

Planet	Bode's Law	Actual Distance
Mercury	4	3.9
Venus	7	7.2
Earth	10	10.0
Mars	16	15.2
—	28	—
Jupiter	52	52.0
Saturn	100	95.0

Although Laplace despised this as a mere number's game, Bode's faith in it seemed to be vindicated when the next term of the series, 192 + 4 = 196, proved to match closely the actual distance of Uranus. Herschel's discovery was hailed as a striking confirmation of the "law."

Bode himself was convinced the gap at number 28 must correspond to a missing planet. Already in 1772, he had written: "Can one believe that the Founder of the Universe had left this space empty? Certainly not." The discovery of Uranus gave impetus to an actual search for the missing body. One of the first to grasp the idea was the Hungarian-born Baron Franz Xavier von Zach.[22] Von Zach was born in Pest (now part of Budapest) in 1754. In 1783 he moved to London, where he became acquainted with Herschel and Count Moritz von Brühl, ambassador of Saxony, who in turn introduced him to Ernst II, Duke of Saxe-Gotha. The duke was interested in setting up a private observatory on the mountain of Seeberg, near his castle at Gotha, and von Zach agreed to become its director. Actual construction got underway in 1788. Meanwhile, von Zach had learned of Bode's prediction that an undiscovered planet lay between Mars and Jupiter. He went so far as to attempt to work out its orbit, but lacked the one element that would point to its location—its longitude on the heavens. Thus he had no choice but to laboriously sift for it among the stars of the zodiac.

After several years of working at his "revision of the stars of the zodiac," von Zach decided the project was too great for a lone observer. In August 1798, during a meeting at Gotha—both Bode and Lalande were in attendance—he mooted the idea of a cooperative venture involving several astronomers. The proposal was premature, but despite the fact that Europe was now torn by warfare, von Zach organized another conference in September 1800 at Johann Hieronymus Schroeter's private observatory at Lilienthal, near Bremen.

Lilienthal was then part of Hanover, and Schroeter's observatory, the grandest on the continent, was partly subsidized by George III.[23] The largest telescope was a 27-foot reflector, surpassed only by William Herschel's 40-foot instrument at Slough. Lalande and other astronomers of France were barred from attending the meeting, since the Hanoverian government by then feared any contact with the radical ideas of the French Revolution.

Apart from von Zach, the others who attended the Lilienthal meeting were Karl Ludwig Harding, Schroeter's assistant; Heinrich Wilhelm Matthäus Olbers, a Bremen physician and amateur astronomer expert in the computation of comet orbits; Johann Gildemeister, also of Bremen; and Ferdinand Adolf von Ende, of Celle. The meeting was briefly interrupted by Adolf Friedrich, prince of Great Britain and Hanover, and his entourage; they were touring the village, especially its newly built bath houses, and wanted to see Schroeter's observatory. The following day, September 21, the group got down to serious business. They discussed von Zach's plan of organizing a systematic search of the zodiac for the missing planet between Mars and Jupiter. The discussions were intense and led to the formation of the "Celestial Police," a company of 24 astronomers who were to participate in the search. Schroeter was chosen president, and von Zach agreed to serve as secretary. The plan called for the constellations of the zodiac to be divided into search zones covering 15° of longitude and 7° or 8 ° north and south of the ecliptic. Each member was to be responsible for scrutinizing the stars of a zone down to the ninth magnitude.

Such plans take time to implement. Before the "Celestial Police" could get underway, surprising news arrived. The missing planet had already been found, but through no systematic plan of search; rather, as with Herschel's discovery of Uranus, it had turned up during the diligent though otherwise directed labors of a Sicilian monk, Giuseppe Piazzi (Figure 7). Ironically, Olbers had written to Piazzi to invite him to join the coordinated effort, though apparently his letter had not yet arrived.

Ever since 1792 Piazzi had been patiently revising the star catalog of the French astronomer Nicolas-Louis de Lacaille. On January 1, 1801, the first night of the new century, Piazzi was hunting down one of Lacaille's stars, of seventh magnitude, in the constellation Taurus. He found the star he was seeking. He also found another, of the eighth magnitude, that did not appear in the catalog. The following night he returned to this position only to find the latter star had slightly shifted its position, moving some 4 arc minutes in right ascension and 3½ arc minutes in declination. For the following nights, he continued to stalk the moving point of light. In letters to Bode and Lalande he cautiously referred to it as a comet, "without a tail or envelope." But to his friend Barnaba Oriani, director of the Brera Observatory of Milan, he confided suspicions of greater consequence: "it might," he allowed, "be something better than a comet." He continued to observe until February 11, when illness forced a temporary retreat from his telescope. By the time he was well enough to return to it, the object had disappeared into the glare of the Sun.

7. Giuseppe Piazzi (1746–1826). (Courtesy of the Mary Lea Shane Archives of the Lick Observatory.)

No one else had been able to observe it, as the news of its discovery had been delayed by the slow mails of a continent at war. However, Piazzi had made enough observations to determine its distance: 2.8 AU. Bode announced that his prediction had been marvelously fulfilled; the trans-martian planet of theory had at last been discovered.[24] Von Zach, too, was convinced and published an article, "On a long supposed, now probably discovered, new major planet of our solar system between Mars and Jupiter."[25]

Needless to say, the new planet's emergence from the Sun in July was eagerly awaited; however, since Piazzi had followed it through an arc of sky of only 3°, predictions of its position were highly uncertain. It was diligently searched for by William Herschel, among others, but without success. In retrospect, we know they were misled by their expectation that the planet would, like Uranus, offer a discernible disk. For awhile it began to seem the planet had been irretrievably lost. Fortunately the Brunswick mathematician Karl Friedrich Gauss, then 24 and at the beginning of his brilliant career, was working on a new method of calculating orbits from only a few observations. Using his method Gauss produced improved ephemerides that led von Zach and Olbers to their quarry on New Year's Eve 1801. Within a few hours of one another, they independently recovered the planet only half a degree from the position Gauss had predicted.[26]

Piazzi named the planet Ceres–Ferdinandea to honor jointly the patron goddess of Sicily and its Bourbon monarch. But the Bourbons were not then held in universal esteem (in France, for instance, the Bourbon king Louis XVI had been guillotined during the Revolution), and the name was quickly abbreviated to Ceres. At first it was assumed to be a new major planet but Herschel's and Schroeter's measures proved otherwise. Instead it was a small body, according to Herschel, only 260 kilometers across. Though this was admittedly an underestimate—the currently accepted value is just under 1,000 kilometers—it was little more than a shard by usual planetary standards.

Tracking Ceres after its rediscovery, Olbers, on March 2, 1802, chanced upon a second body, Pallas. It, too, was moving in the "gap" between Mars and Jupiter. Olbers pondered the implications in a letter to Herschel. "Could it be," he wrote, "that Ceres and Pallas are just a pair of fragments, or portions of a once greater planet which at one time occupied its proper place between Mars and Jupiter, and was in size more analogous to the other planets, and perhaps millions of years ago, either through the impact of a comet, or from an internal explosion, burst into pieces?"[27] If so, there was every reason to expect that diligent searching would turn up still more of these planetary disjecta membrae, for which William Herschel coined the not wholly suitable term "asteroids" (referring to their starlike telescopic appearance; but of course there is nothing really starlike about them). Rather than disbanding in the aftermath of the discovery of Ceres, the "Celestial Police" decided to keep up their quest.

Their efforts were duly rewarded. Harding, on September 1, 1804, captured a third asteroid, Juno. Olbers, on March 29, 1807, added a fourth, Vesta. In 1816 Olbers, who alone of the group had kept the enterprise alive, abandoned the quest; he had decided that further efforts were futile.

There was a hiatus until 1830, when Karl Ludwig Hencke, postmaster of Driessen, resumed the search. For 15 years he worked alone and unaided. In December 1845, he discovered the small planet Astraea. Two years later he added another, Hebe. Now discovery began to follow discovery. By 1851, 15 asteroids were known, and George Biddell Airy, Astronomer Royal of England, predicted: "We shall go on finding more planets, till space, in our solar system at least, may appear, in reference to what was formerly known, comparatively full of these travelling bodies."[28] Scores more were added by diligent observers such as Hind, de Gasparis, Luther, Goldschmidt, and Charcornac. Not a one of these is a household name today, but in their time they were trailblazers into unknown interplanetary space. Some were professional astronomers, others amateurs. Goldschmidt, for instance, earned his living as an artist.[29]

During the years Olbers, and later Hencke, carefully searched trans-martian space, a larger game was afoot. In the years after 1781, a faint but perceptible pulse announced the possible existence of a more remote but far grander body. It was registered by observers intent on following Uranus along the widening arc of its orbit and was to lead to one of the most spectacular events in the history of astronomy, indeed, the very "Zenith of Newtonian mechanics." It would render the names of Urbain Jean Joseph Le Verrier and John Couch Adams immortal among planet-seekers.

six

Le Maître Mathématicien

The discovery of Uranus had destroyed forever the ancien régime of the heavens, the complacent view from time immemorial that Saturn marked the ne plus ultra of the solar system. In terrestrial affairs, too, change was in the air. In France, the old order, developed by imperceptible degrees over millenia and persisting ever since feudal times, was in disarray. The chaos and upheaval of the French Revolution and the beheading of France's Bourbon King Louis XVI sent shock waves through the rest of Europe. During the ensuing terror, no man of ability or ambition was safe. Jean Sylvain Bailly, distinguished historian of astronomy, was guillotined. So was Jean Baptiste de Saron, the first man to calculate an orbit for Uranus—he spent the night before his execution calmly calculating a refined orbit for Halley's Comet. Lagrange and Laplace were spared, possibly because they were requisitioned to calculate trajectories for the artillery and to help direct the manufacture of saltpeter for gunpowder. Laplace, nevertheless, was viewed with suspicion; he was removed from the Commission on Weights and Measures because he was thought to be lacking "republican virtues and the hatred of kings." He took refuge at Melun, southeast of Paris, where he wrote his *Exposition du système du monde* (published in 1796). The first edition was dedicated to the Council of the Five Hundred, with a suitably egalitarian preface:

> The greatest benefit of the astronomical sciences is to have dissipated errors born of ignorance of our true relations with nature, errors all the more fatal since the social order must rest solely on these relations. Truth and justice are its immutable bases. Far from us be the dangerous maxim that it may sometimes be useful to deceive or to enslave men the better to insure their happiness! Fatal experiences have proved in all ages that these sacred laws are never infringed with impunity.[1]

When Napoleon came to power, Laplace, who had examined and passed the future First Consul as a 16-year-old student at the École Militaire, was appointed head of the Ministry of the Interior in 1799. Napoleon found him "a mediocre administrator" and removed him in favor of his brother Lucien Bonaparte after only 6 weeks. But Laplace remained in favor as a mathematician and received all the Napoleonic orders of note, including the Grand Cross of the Legion of Honor. After Napoleon's defeat in 1814, Laplace, as a member of the Senate, voted for the end of Napoleonic power and did not rally to the emperor's cause during his brief return from Elba (the "Hundred Days" from March 1815 until Waterloo, June 18, 1815). Thus he remained in favor with Louis XVIII after the Bourbon restoration and received a seat in the Chamber of Peers. Now as the Marquise de Laplace, he wrote another, equally noble preface for his book:

> Let us conserve with care and increase the store of this advanced knowledge, the delight of thinking beings. It has rendered important services to navigation and geography; but its greatest benefit is to have dissipated the fears produced by celestial phenomena and to have destroyed the errors born of ignorance of our true relations with nature, errors which will soon reappear if the torch of the sciences is extinguished.[2]

During these years of violent change, Laplace continued to march along the path he chose at the beginning of his career. "To all his works," wrote Fourier, "he gave a fixed direction from which he never deviated; the imperturbable constancy of his views was always the principal feature of his genius. . . . He had solved a capital problem of astronomy, and he decided to devote all his talents to mathematical astronomy, which he was destined to perfect. He meditated profoundly on his great project and passed his whole life perfecting it with a perseverance unique in the history of science. The vastness of the subject flattered the just pride of his genius."[3] The result was the sublime *Mécanique Céleste*. The first two volumes appeared in 1799, one volume each came out in 1802 and 1805, and a fifth and last volume was published in 1825.

The *Mécanique Céleste* is a difficult work, not the least because of Laplace's penchant, in condensing his arguments, for skipping over numerous intermediate steps. Often, when Biot was assisting him in revising it for the press, Laplace himself could not recover the details. So long as he was able to satisfy himself his conclusion was correct, the intermediate steps did not matter to him, and he simply replaced them with the phrase *"Il est aisé à voir"* (It is easy to see); he had a reputation for being vain and selfish and for borrowing heavily from others without attribution.

For all his shortcomings, Laplace's vision was consistent and sublime; he kept constantly before him a lofty conception of the eternally stable solar system repeating forever the complicated cycle of its motions. In marked contrast to the vicissitudes of human affairs, the heavens seemed to present a consoling image of predictable order. As earlier noted, Newton had glimpsed the unseemly tendency of planetary perturbations to accumulate to the point where they might even destroy the order of the solar system. He had even suggested God might need to intervene from time to time to set it right again. Laplace was an atheist; his massive attempt to work out the stability of the solar system was largely an attempt to remove the last obstacles to the triumph of reason. His final system, indeed, was godless. Thus his famous retort to Napoleon, when the latter asked why in all the *Mécanique Céleste* the consummate mathematician had not once mentioned the Creator: "Sire, I had no need of that *hypothesis!*"[4]

The challenge for Laplace and others who devoted themselves to the remote and exacting specialty of celestial mechanics was to narrow as far as possible the gap between theory and observation, until the differences amounted to less than the errors of the observed positions. The perturbation theory developed for this heroic enterprise was a march into the unknown country of theory, the blazing of a trail through difficult uncharted territory. Workers were driven by actual problems and the obsessive question: Is it possible to calculate every motion by means of Newton's theory?

According to an elegant mathematical theory created by Lagrange in 1774, the mutual perturbations of the planets could be expressed in terms of a "perturbing function," defining the cross-pulling and hauling of the planets in terms of their distances. In order to be grappled with practically, this function had to be broken down into individual components, which were expressed as series expansions dependent on the eccentricities and inclinations. The complications emerged quickly by virtue of the fact that higher-order terms, though minor in the original expansion, often became more and more significant with successive integrations—the important lesson Clairaut had learned in the case of the advance of the Moon's apsides.

Laplace himself developed the perturbative function to include the third-order terms of the eccentricities and inclinations. In taking still higher (fourth-order and above) terms into account, "the labor of computing the coefficients increases with frightful rapidity,"[5] to the point where all hopes of determining them by the ordinary process of algebraic development became so complicated as to be well nigh unmanageable.

The classic three-body problem, the Moon's motion, had by now become a specialized area of research unto itself. "The exact solution," wrote Laplace, "surpasses the powers of analysis; but from the proximity of the Moon, compared with its distance from the Sun, and from the comparative smallness of its mass, an approximation may be obtained extremely near the truth. Nevertheless, the most delicate analysis is necessary to extricate all the terms, whose influence becomes sensible."[6]

Planetary theory, the calculation of the effects of the mutual perturbations of the planets, was, of course, quite as complicated as lunar theory. It was largely the domain of the French, who attempted to extend the problems and methods defined by Laplace. Laplace, no innovator or paradigm-shifter himself, but the great "hero of normal science,"[7] had "definitively placed the law of universal gravitation in the rank of the great scientific truths by showing that the phenomena that appeared to contradict it or limit its domain," such as the great inequality of Jupiter and Saturn and the secular acceleration of the Moon, were explained by it.[8] He died in 1827, a century after Newton. His final words were "life is short, knowledge is vast," in emulation of Newton's "I do not know what I appear to the world; but to myself I seem to have been only like a boy playing on the seashore." By then it was generally accepted Newtonian theory could explain away every nod and wiggle in the motions of the Moon, planets, their satellites, and the ephemeral comets. The grand program as defined by Laplace was as follows:

> If the planets only obeyed the action of the Sun, they would revolve round it in elliptic orbits, but they act mutually upon each other and upon the Sun, and from these various attractions, there result perturbations in their elliptic motions, which are to a certain degree perceived by observation, and which it is necessary to determine to have exact tables of the planetary motions. The rigorous solution of this problem, surpasses the actual powers of analysis, and we are obliged to have recourse to approximations. Fortunately, the smallness of the masses compared to that of the Sun, and the smallness of the eccentricity and mutual inclination of their orbits, afford considerable facilities for this object.[9]

He had expressed his faith that "the law of universal gravitation . . . represents all the celestial phenomena even in their minutest details."[10] It was very much a faith. Though straightforward in principle, the application of perturbation theory was, in practice, a hideous matter. The increasingly complicated equations used in the calculations were becoming, as historian Watson Warren Zachary has noted, "as far removed from the inverse square law as epicycles on epicycles . . . were removed from uniform circular motion."[11]

Astronomy, as Laplace had defined it, was reduced to a grand problem in mechanics. As such, it had become the science of obtaining precise positions, duly corrected for all conceivable distortions, such as refraction due to the atmosphere, temperature, and barometric pressure. It was also a matter of indefatigable calculation to bring these reduced observations into agreement with the demands of theory. The background stars were, in terms of these preoccupations, of interest only insofar as their positions provided a fixed reference framework against which the motions of the Moon and planets could be measured.

The rationalist faith that the Newtonian theory represented "impregnably the system of the world and [was] beyond reproach"[12] would appeal mind and soul to one Urbain Jean Joseph Le Verrier (Figure 8). Like his great hero Laplace, Le Verrier was a native of Normandy. He was born at St. Lô (département de la Manche) on March 11, 1811. The region had been one of those to strongly resist the Revolution. In politics Le Verrier was steadfastly conservative and monarchistic. His father was a bureaucrat, an employee of the State Property Administration, whose ambitions for his son were only exceeded by the son's ambitions for himself.[13] Le Verrier received 2 years of education in mathematics at Caen, where Laplace had studied for 5 years. He graduated at the head of his class. However, his provincial education was not adequate to prepare him for the difficult entrance examination to the prestigious École Polytechnique in Paris. He was not admitted. His father promptly sold his house to secure the funds needed to send him to the Collége de Saint Louis in Paris, where Le Verrier could further prepare himself.

Le Verrier took advantage of the opportunity, and duly gained admission to the École. One can picture him as a student there: handsome features, cool and penetrating eyes, an aloof and aristocratic bearing. He impressed his peers, but not unusually so. One of them commented that though his performance seemed "solid" and gave "promise of an honorable career, no one could then have predicted his future brilliance."[14]

Nevertheless, he earned honors, and toward the end of his connection with the École (or immediately thereafter) he provided George Biddell Airy, Plumian professor of astronomy at Cambridge, with his analysis of the motions of the Earth and Venus, a problem with which Airy was then engaged. Airy, indeed, always considered himself Le Verrier's "oldest scientific friend," and dated their connection from this time.

8. U. J. J. Le Verrier (1811–1877). (Courtesy of the Yerkes Observatory.)

Meanwhile France was again beset by political change. Louis XVIII had died, replaced by the reactionary Charles X. In the July Revolution of 1830, Charles had been overthrown and replaced by the bourgeois régime of King Louis Philippe of Bourbon-Orléans. It was one of Louis Philippe's policies to offer the graduates of the École Polytechnique their choice of positions in the departments of public service. The recently graduated Le Verrier, lured by the great name of Gay-Lussac, chose to work as an experimental chemist in the Administration des Tabacs. As always, he worked hard, publishing two creditable papers on the chemistry of phosphorus.

Meanwhile he had been drawn into the brilliant social life of Paris, where he was frequently to be found in the company of one Mlle. Choquet. Since a requirement of Louis Philippe's civil servants was to carry out field work in the outlying provinces, Le Verrier in 1836 faced a difficult decision: leave Paris or resign his position. He chose the latter, and spent a year as a tutor at the second-rate Collége Stanislaus. He also married Mlle. Choquet.

By 1837, two positions had opened up in the École Polytechnique: one of répétiteur (assistant) to Gay-Lussac in chemistry, the other of répétiteur in astronomy. Undoubtedly Le Verrier would have preferred to stay in chemistry, but Gay-Lussac had to choose between him and another excellent candidate, Henri Victor Regnault. Knowing of Le Verrier's interest in working out advanced problems in celestial mechanics, Gay-Lussac resolved his dilemma by proposing Regnault for the position in chemistry, Le Verrier for that in astronomy. Obviously he judged correctly; Le Verrier, "without regret as without effort, without dividing his attention and without looking back, detached himself from chemistry and, obedient to the decree of chance that pointed out his course, rapidly became an astronomer."[15] Within a few days of this sudden career diversion, he wrote to his father, "I must not merely accept, but must actively seek out opportunities to improve my knowledge." He was, moreover, supremely confident of himself. "I have already begun to mount the ladder [of success]," he added; "why shouldn't I continue to climb?"[16]

Now an astronomer, he set his sights high. His first choice of an essay was "Sur les variations seculaires des orbites des planètes" (On the secular variations of the orbits of the planets). The problem was fundamental to the question of the stability of the solar system and had been the subject of a profound memoir by Lagrange, published in 1782. However, the latter's results had long been suspect because of the erroneous values he adopted for the smaller planets (his mass for Venus, for instance, had been 50 percent too great). Le Verrier proceeded to calculate, at 10,000-year intervals, all the elements of the orbits of Mercury, Venus, Earth, and Mars for 100,000 years before and after epoch 1800. His results agreed very closely with those of Lagrange, although the orbital elements adopted by the two mathematicians were very different. It turned out the erroneous mass of Venus adopted by Lagrange had fortuitously canceled out its own error. Le Verrier's paper was delivered to the Paris Académie des Sciences on September 16, 1839, to favorable comment.[17] But important as its particular results were, the fact a vast subject had been thrown open to him was even more so. Le Verrier had made a fateful choice, and it determined the direction of all his future work: "This proved to be the beginning of a colossal undertaking, which, if considered at first in its entirety, would have appeared to most persons to be one impossible to be completed in the limited span of a single life. Le Verrier himself probably hardly intended to include in his investigations the theories of all the major planets."[18]

Over the next few years, Le Verrier published several investigations concerned with the long-term oscillations of the orbital elements of the planets—work that attracted the notice of the doyen of French astronomers, François Jean Dominique Arago, director of the Paris Observatory since 1830. In 1841, Arago suggested Le Verrier apply his unique analytical skills and almost superhuman endurance for calculation to master the motion of Mercury. This planet was always the most irksome, partly because of its proximity to the Sun, which made exact determinations of its position difficult, and partly because the magnitude of the eccentricity of its orbit (0.205, the largest of all the planets then known) multiplied the effects of these errors. The problem of Mercury had, as we have seen, baffled astronomers ever since the time of Ptolemy. Yet Le Verrier, who was always ready to please a superior, did not hesitate. "Docile to circumstance," he set aside all other work and threw himself wholeheartedly into the theory of Mercury.

He approached his task with characteristic resoluteness. A master planetary accountant, he set about reducing all available observations in order to square every last jot and tittle of the planetary account books. Mercury's motion was, indeed, a critical test of gravitational theory, since "if the theory accounted for the perturbations, then it followed that tables could be constructed, based on the theory. So, the ultimate test of Newtonian gravitational mechanics boiled down to the agreement between the predicted tabular position and that actually observed."[19] This agreement became the grail Le Verrier was seeking.

Le Verrier used two sets of observations in his work on Mercury. To the first set belonged 400 observations of Mercury that had been taken with the meridian-circle at the Paris Observatory between 1801 and 1842. In general these agreed with the calculations to within a few arc seconds, perhaps an acceptable level of accord given the extreme difficulties of obtaining precise positions of the planet. To the second set belonged the timings of the contact-points (the points of entry or exit of Mercury on the Sun's disk) made at the 12 best observed transits of Mercury between 1697 and 1832. By their very nature, the latter points were fewer in number but much more accurate than the meridian-circle observations.

Before Le Verrier, attempts to calculate tables of Mercury had been notoriously unsuccessful. Predictions for the 1707 transit were 1 day in error; the tables of La Hire and Halley missed that of 1753 by many hours, while Lalande's tables had failed by 53 minutes in 1786, with the consequence that all the astronomers of Paris except Delambre had abandoned their telescopes prematurely. These experiences created fresh alarm and redoubled efforts. Lalande published improved tables for the transits of 1789, 1799, and 1802, while Lindenau's tables of 1813 enjoyed a brief vogue. Neither, however, could reconcile the remaining discrepancies.

Each calculator had found it easy to discover blemishes in the work of his predecessor. Lalande had criticized the supposed sloppiness of Halley, only to be embarrassed by his own tables. In like manner, Le Verrier, who seems to have taken particular relish in scoring the faults of others, took Lindenau to task for having made elementary mistakes of arithmetic. Le Verrier himself struggled with the intractable problem for 3 years, publishing a paper in 1843, "Détermination nouvelle de l'orbite de Mercure et de ses perturbations." Two years later he followed it up with a book, *Théorie du Mouvement de Mercure*. Although in his paper he had expressed reservations about subduing the errant planet without recasting the tables of the Sun's apparent motion (and, since the apparent motion of the Sun was a reflection of the Earth's motion, the motion of the Earth itself), he actually suppressed them from the book. There he gave a prediction for the transit of May 8, 1845, which would be partially visible from France and entirely from North America. Perhaps he was merely hedging his bet. As Zachary surmises, he may have felt that if his tables were accurate, "so much the better for him; if they failed," he could advert to the difficulties encountered by his predecessors.[20] In any case the stakes were high at the 1845 transit.

The event was eagerly awaited by Ormsby Macknight Mitchel. He was ready with the 11-inch Merz equatorial of the Cincinnati Observatory, a glass, he declared, "in search of which I had traversed the ocean and the land." The cornerstone for the observatory had been laid 2 years earlier by former United States president, John Quincy Adams. As Mitchel awaited the transit, his excitement rose to the highest possible pitch:

> I had the high satisfaction of seeing mounted one of the largest and most perfect instruments in the world. I had arranged and adjusted its complex machinery, had computed the exact point on the sun's disc where this planet ought to make its first contact, had determined the instant of contact by the old tables, and by the new ones of Le Verrier, and, with feelings which must be experienced to be realized, I took my post at the telescope to watch the coming of the expected planet. After waiting what seemed almost an age, ... I caught the dark break which the black body of the planet made on the bright disk of the Sun. 'Now!' I exclaimed; and, within *16 seconds* of the computed time, did the planet touch the solar disc, at the precise point at which theory had indicated the first contact would occur.[21]

Another observer whose attention was fixed on the Sun that fateful day in May was Edmond Modeste Lescarbault, a physician and amateur astronomer of Orgères-en-Beauce, in the arrondissement of Chateaudun, Department of Eure et Loire. Interested in astronomy since boyhood, he had long entertained a particular hobby-horse: the existence of unknown planets. For some years this was not a well-developed notion, only an intimation from his having noticed in 1837 that Bode's law was not quite accurate in predicting the distances of the planets from the Sun. From this he speculated, in addition to the four small planets between Mars and Jupiter, another might be lurking elsewhere in the system. At the moment he undertook nothing so systematic as the lonely reconnaissance that engaged Hencke, which in December 1845 had led to the recognition of Astraea. However, his idea was not forgotten.

The impression made on Lescarbault by the transit of Mercury in 1845 stimulated a new thought: Suppose apart from Mercury and Venus, another small planet lay between the Earth and the Sun. It must occasionally transit the solar disk, and might be recognized then. Improbable as the idea might seem, there was certainly no a priori reason to discount it, nor was Lescarbault alone in this reflection. In 1826, a Dessau pharmacist, Heinrich Schwabe, began a systematic search for intramercurial planets. Although he found no planet, his careful scrutiny of the spots on the solar disk led to the unexpected disclosure of the 11-year sunspot cycle. Two other Germans, J. F. Benzenberg and his assistant J. F. J. Schmidt (who later gained fame for his great map of the moon), had scrutinized the solar disk with the same goal since 1834. For that matter, the archives contained many reports of what might be regarded as unscheduled transits by suspicious bodies. None of them really commanded belief, yet they did introduce a modicum of doubt about the satisfactoriness of existing knowledge. They hinted there might be undiscovered country awaiting exploration.

For now the doctor of Orgères slips quietly from our story, but the arresting image of Mercury on the face of the Sun had made its impression, and Lescarbault will reappear again, in dramatic fashion, a few years hence.

The 16 seconds error in Le Verrier's calculation of the May 8, 1845, transit, so slight as to seem almost uncanny to Mitchel, was nevertheless more than could be tolerated—enough to shatter Le Verrier's confidence in his result. Le Verrier knew this and immediately stopped publication of his tables of Mercury, which were even then being typeset by the Bureau des Longitudes. Not yet ready to begin the wider investigations involving the Sun's tables he had mooted in his 1843 paper, he beat a temporary retreat from the toils of planetary theory.

Le Verrier did not, of course, waver in the truth of gravitational theory; his faith in it remained absolute. At the moment, however, he left the motion of Mercury as unfinished business and turned from the seemingly impossible to the merely difficult, working out, brilliantly, orbits for two new comets.[22]

While still absorbed in these researches, Arago again tapped him on the shoulder. Another planetary problem had been disturbing the sleep of the illustrious director of the Paris Observatory. It was a problem as intractable as that of Mercury and required the same judgment, thoroughness, and analytical skill. Herschel's planet, Uranus, was off course. Arago put to Le Verrier the vexsome problem, and as before, Le Verrier proved "docile to circumstance." Without preconception other than the impregnability of the law of gravitation and the power of his own mathematics, he set to work tracing in detail the digressions of the outermost planet. Once again he found himself in a trail-blazing circumstance, although this time, unbeknownst to him, he was not alone.

seven

Le Verrier's Planet

The motion of Uranus had already nagged at astronomers for decades, and by the mid-1840s had become, if not a crisis of gravitational theory, at least "a worrisome anomaly at the very heart of astronomy."[1] The nature of the problem had begun to emerge as early as May 1782, when Lalande noted that the planet newly discovered by Herschel did not appear to be strictly adhering to its prescribed path. At the time, this concern caused no great measure of alarm. Father Placidus Fixlmillner, using the 1690 observation of Flamsteed and that of 1756 by Mayer to backtrack the planet over a larger arc of its path, calculated an orbit and tables of its motion in 1784. Four years later, his tables no longer seemed to suffice. Fixlmillner then attempted to recalculate the planet's orbit, and in doing so realized he could devise orbits that would satisfy either the old observations or the recent ones made since 1781, but not both. Under the circumstances, he decided to rely only on the recent data.

Several other astronomers, including Oriani and Delambre, attempted to remedy the situation by taking more accurate account of the perturbations of Jupiter and Saturn on Uranus. Delambre's efforts seemed to be crowned with success in 1790, when he published tables in tolerable agreement with the observations made since 1781, as well as with the two "ancient" observations then known (Flamsteed's position of 1690 and Tobias Mayer's of 1756). But by 1800, it was evident even Delambre's tables would no longer do—Uranus, it seemed, was veering inexplicably off course. Lalande and Johann Karl Burckhardt, a leading German computer of orbits, attempted to resolve the enigma in a novel manner; they "concluded that there existed an unseen planet beyond Uranus, and they occupied themselves in trying to discover its position."[2] However, they kept the matter strictly between themselves. The question indeed was still in a premature state of development and not far enough advanced to lead to definite results. Lalande and Burckhardt had had a keen insight, but they consulted no one and influenced no one.

The next person to grapple with the problem of Uranus was a French astronomer, Alexis Bouvard. He had been a shepherd in the village of Chamonix, in the valley below Mont Blanc. At 18 he set out for Paris, discovered a talent for mathematics, and won the patronage of Laplace, for whom he worked out the detailed calculations of the *Mécanique Céleste*. Later he was attached to the Paris Observatory, and for many years supervised the calculation of the ephemerides of the *Annuaire du Bureau des Longitudes*. His tables of Jupiter and Saturn appeared in 1808—their accuracy marred by considerable errors in the masses adopted for the two planets. In 1820, he set out to correct these errors, and also those in Delambre's tables of Uranus.

When Bouvard began this immense labor, he was 53. By then, Uranus's motion had been covered by modern observations for nearly 4 decades—almost half an orbit. Moreover, Bouvard had 17 "ancient" observations at his disposal, going back to Flamsteed's observation of 1690. Bouvard himself had published no less than 12, from the records of his deceased colleague, Pierre Lemonnier. He implied, however, that Lemonnier had recognized only 3 of these observations, and that he, Bouvard, had discovered the rest. In reality, the reverse was true. Unfortunately, this was not Bouvard's only—or most serious—misrepresentation of Lemonnier's work.[3]

With such records at his command, Bouvard had every reason to hope for success. And yet he tried to force them; he simply could not make all the observations fit. His solution was to throw out those made before 1781. By doing so, he was able to finish his calculations. In 1821, he published his revised tables of Jupiter, Saturn, and Uranus. However, he entered a caveat in the introduction:

> The construction of the tables of Uranus involves this alternative: if we combine the ancient observations with the modern ones, the first will be adequately represented, but the second will not be described within their known precise tolerances; while if we reject the ancient positions and retain only modern observations, the resulting tables will accurately represent the latter, but will not satisfy the old figures. We must choose between the courses. I have adopted the second as combining the most probabilities in favor of truth, and I leave to the future the task of discovering whether the difficulty of reconciling the two systems results from the inaccuracy of the ancient observations, or whether it depends on some extraneous and unknown influence which may have acted on the planet.[4]

Bouvard hereby disclosed an unpleasant dilemma. In throwing out the old observations he was, in effect, as Morton Grosser notes, charging four first-rate observers with incompetence.[5] Bouvard attributed an error of 65".9 to Flamsteed's observation of 1690, despite the fact his observations could generally be relied upon to 10". The observations of Bradley, Mayer, and Lemonnier were assumed in error by 40" or 60", again badly off their usual standard of 5" or 6".

Lemonnier, who died in 1799, was the most seriously slandered. It is true Bouvard had strong motives to discredit Lemonnier's positions, and he proceeded as theoreticians often do when there is trouble with their theories, that is, by foisting the blame on the observer (there was a precedent for this, as we have seen, in Newton's attempt to charge his failure with the lunar theory to Flamsteed's account). Lemonnier seems to have been a rather ill-tempered man and is said to have quarreled with nearly everyone he met. Yet he was a careful observer for all that, totally dedicated to his work of watching the stars as they crossed the five vertical wires of his transit instrument and recording their transit times in the large volumes of his observational log. His reputation, however, is otherwise; he has been characterized as a bumbler who through carelessness deprived himself and his countrymen of a great discovery. This accusation first appears in François Arago's *Astronomie Populaire* and has been endlessly repeated ever since. Arago claimed Lemonnier's records presented "an image of chaos," implying they might contain gross errors. Arago believed one of Lemonnier's observations of Uranus had been hastily scrawled on a bag that once contained hair perfume (thereby linking them to the powdered wig, a hated symbol, for an ardent republican like Arago, of the ancien régime)! However, if one of Lemonnier's observations was ever recorded on a paper bag, it was most assuredly not one of the Uranus observations, as Dennis Rawlins has shown conclusively from an inspection of the actual records in the Paris Observatory. They were all recorded in the usual fashion in Lemonnier's observing log. Lemonnier did record the planet eight times in less than a month from December 1768 to January 1769—including four successive nights between January 20 and 23—but his failure to recognize the planet was hardly owing to carelessness, rather to sheer ill luck. His telescope was small since it was designed for transit work, and the planet would not have shown a perceptible disk such as would later betray its unusual nature to Herschel. The only way Lemonnier could possibly have recognized it as a planet would have been by its motion. At the time, the planet happened to be near its stationary point—thus it was *not* moving. It is hardly surprising Lemonnier, with other purposes in view, failed to recognize the planet. Therefore, his good name deserves to be fully rehabilitated. (We have already mentioned this once, but since the slander has been so often repeated, we feel obliged to mention it again!) Alas, as we shall see, his was not the only reputation to be wrecked on the problem of Uranus.

Bouvard's trashing of Lemonnier served no purpose. Even without the old observations, the best orbit he could manage still contained unacceptably large errors of 9 arc seconds. Moreover, he had made mistakes in the calculations. Some of these were pointed out by the German astronomer Friedrich Wilhelm Bessel, who early in his career had received training at Schroeter's observatory at Lilienthal and was now director of the observatory at Königsberg,

Prussia. Bessel, perhaps because he was an observer as well as a the-
oretician, had more faith in the competence of the old observers
than Bouvard did. He began to suspect the discrepancy in Uranus's
motion might lie with the simple inverse-square law of gravitation
itself.[6] In any case, the discrepancy was quite real; it could not be
disguised by mere sleight-of-hand. "I have myself . . . attained the
strong conviction that the existing differences . . . are by no means
attributable to the observations," he wrote.[7]

During the early 1820s, the errors in Bouvard's tables continued to mount, and it was clear future errors would be at least as
large, if not larger than those of the past. Briefly, in 1829–1830, the
tabular and observed longitudes coincided, but the respite was
short-lived. Afterwards the planet began to lag behind its calculated place.

Taking stock of the situation in 1832, George Biddell Airy,
Plumian professor of astronomy at Cambridge (Figure 9), reflected in his "Report on the Progress of Astronomy" to the British
Association for the Advancement of Science that there was no obvious way of squaring the pre- and post-1781 observations. When
Airy wrote, the errors in longitude of Bouvard's *Tables* had again
accumulated to an intolerable .5 arc minutes; and were still increasing. Airy concluded his discussion by saying, "We have no
clue to their explanation."[8]

Was the planet impeded by a resisting fluid or "thin ethereal
medium" of some kind?[9] Was it perturbed by a large satellite, or
had it been struck by a comet at about the time it had been discovered, so that its orbit had been radically changed? (This was a
desperate attempt, clearly, to reconcile the pre- and post-1781 observations, an attempt that crumbled to nought when the planet
continued to defy Bouvard's orbit, which had been worked out exclusively on the basis of the latter observations.)

None of these theories was ever convincing. By the time Airy
wrote, the possible solutions had narrowed to two. One was the
alternative Bessel had meditated—of adjusting the inverse-square
law, as historian Robert W. Smith recently observed.[10] But Newton's law was not to be lightly set aside. Although there was a long
tradition of doubt reaching back to Clairaut and Euler, the law had

9. George Biddell Airy (1801–1892). (Courtesy of the Yerkes Observatory.)

always emerged triumphant. For this reason, attention was focused increasingly on another possibility: Something was drawing Uranus off course.

Lalande and Burckhardt had apparently swum in these waters already, only, one suspects, to turn back when the practical difficulties began to swarm around them. Bouvard, too, in his long meditations on the motion of Uranus had sensed something more definite than the shadowy "extraneous and unknown influence" mentioned in the introduction to his *Tables*—nothing less than another planet exterior to Uranus, which was perturbing the latter's motion. He had corresponded on the subject with Peter Andreas Hansen at the Seeberg Observatory and had also discussed it with an Englishman visiting Paris, Rev. T. J. Hussey, Rector of Hayes in

Kent. For that matter, the exterior planet idea had independently dawned upon Hussey, and when he discussed it with Bouvard the response had been encouraging—enough for Hussey to trouble Airy with a note on November 17, 1834:

> The apparently inexplicable discrepancies between the ancient and modern observations suggested to me the possibility of some disturbing body beyond *Uranus*, not taken into account because unknown. My first idea was to ascertain some approximate place of this supposed body empirically, and then with my large reflector set to work to examine all the minute stars thereabout: but I found myself totally inadequate to the former part of the task . . . . I therefore relinquished the matter altogether; but subsequently, in a conversation with Bouvard, I inquired if the above might not be the case: his answer was, that, as might be expected, it had occurred to him. . . . Upon my speaking of obtaining the places empirically, and then sweeping closely for the bodies, he fully acquiesced in the propriety of it, intimating that the previous calculations would be more laborious than difficult; that, if he had the leisure, he would undertake them and transmit the result to me, as the basis for a very close and accurate sweep. . . . If the whole matter do not appear to you a chimera, which, until my conversation with Bouvard I was afraid it might, I shall be very glad of any sort of hint respecting it.[11]

Perhaps Hussey hoped Airy, a first-rate mathematician and former Senior Wrangler in mathematics at Cambridge, would seize the opportunity and carry out the calculations himself; a hint of this was in his letter. If so, Hussey was disappointed. Airy was enough of a mathematician to appreciate the difficulties of what Hussey was proposing, and his icy reply could hardly have been more discouraging:

> It is, a puzzling subject, but I give it as my opinion, without hesitation, that it is not yet in such a state as to give the smallest hope of making out the nature of any external action on the planet. . . . I am sure it could not be done till the nature of the irregularity was well determined from several successive revolutions.[12]

Since Uranus took 84 years to complete one orbit, this came out, roughly, to several centuries! Needless to say, Hussey abandoned all further interest.

Nevertheless, the idea of an exterior planet perturbing Uranus was one whose time had come. The following year Halley's comet made its first return to perihelion since 1759. The magnitude of the refinements that had occurred in celestial mechanics in the meantime and the invincible accuracy of Newton's law were dramatically demonstrated. Whereas Clairaut, Lalande, and Madame Lepaute were acclaimed for calculating the return in 1759 within 33 days, several astronomers pinpointed the 1835 return to within a few days. The most notable was the French astronomer Phillipe Le Douclet comte de Pontécoulant, who took into account the perturbative effects of Jupiter, Saturn, and Uranus between 1682 and 1835 and those of the Earth near the comet's 1759 perihelion passage; he was off by only 3 days in predicting the comet's November return to perihelion. As small as the difference was, it led Jean Élix Benjamin Valz at the Marseilles Observatory and Friedrich Bernhard Gottfried Nicolai at the Mannheim Observatory to declare that some unknown force must have acted to retard its motion. Valz wrote, "I would prefer to have recourse to an invisible planet beyond Uranus. . . . Would it not be admirable thus to ascertain the existence of a body which we cannot even observe?" Nicolai, resorting to Bode's law, gave a still more specific expectation. "One immediately suspects that a transuranian planet (at a radial distance of 38 astronomical units, according to the well-known rule) might be responsible for this phenomenon."[13] The idea of an exterior planet thus moved at once from vague inference to more perceptible and determinate shape.

For that matter, was it possible the offender had already been sighted? The previous May, Niccolo Cacciatore, successor to Piazzi as director of the Palermo Observatory, had come across a moving object of the 8th magnitude. Although horrible weather followed, preventing him from observing the object again before it was lost in the evening twilight, it seemed to be moving slowly enough "to make me suspect the situation to be beyond Uranus."[14] The publication of Cacciatore's tantalizing, but inconclusive, observation led to publication of other observations made during the summer of 1831 by Louis François Wartmann, a

Genevan astronomer. Wartmann's object was, however, in a different part of the heavens, so that it could not have been identical to Cacciatore's. It, too, remained unconfirmed, and for that matter did not inspire much confidence—there was, as John Russell Hind was later to comment, "something strange about the whole matter."[15] Nevertheless, these fugitive sightings lent credence to a growing mood of expectation, a sense of something portentous about to be revealed.

Meanwhile the celestial mechanicians continued to refine their calculations. Alexis Bouvard once more undertook the laborious revision of his tables of Jupiter and Saturn, and—perhaps a sign of his mounting frustration and fatigue—delegated to his nephew, Eugène, the thankless task of reconstructing those of Uranus. Eugène was nothing if not diligent. On October 6, 1837, he wrote inquiringly to Airy, who had recently left Cambridge to become Astronomer Royal at Greenwich, "that the differences in latitude are very large, and are continually becoming larger. Does this indicate an unknown perturbation exercised upon the motions of this star by a body situated beyond? I do not know, but this is at least my uncle's idea."[16] Airy had the information Eugène requested, having made a careful study of the observations of Uranus obtained at Cambridge in 1833, 1834, and 1835 and at Greenwich in 1836. He had, he told Eugène, satisfied himself that the errors of latitude were not increasing, but those of longitude were doing so "with fearful rapidity." The planet was lagging ever farther behind its predicted place.

This, however, was not the only problem. There was (to Airy's mind) an even more consequential error. The radius vector—the imaginary line between the planet and the Sun, and the central quantity in both Kepler's laws and Newton's dynamics—always came out "considerably too small." This was a point on which, then and later, Airy put particular emphasis. The Astronomer Royal concluded his letter to Eugène Bouvard with a deflating statement similar to that he had made to Hussey. Bouvard, he proposed, "would gain much . . . by waiting," since even if the errors were due to an "unseen body, it will be nearly impossible ever to find out its place."[17] Whether this led Bouvard to set aside the planetary hypothesis for the moment and take a more pedestrian approach to the problem is uncertain. In any event, that is exactly what he did.

Others were now being drawn into the problem. According to the American astronomer Benjamin Apthorp Gould, Jr., "numerous mathematicians . . . conceived the purpose of entering into laborious and precise calculations, in order to determine whether the assumption of an exterior cause of disturbance were absolutely necessary, and, if so, to determine from the known perturbations their unknown cause."[18] One was Bessel, who by 1840 had overcome his doubts about the exactness of the Newtonian law and instead sought to ravel back the discrepant motions of Uranus by assuming the existence of an exterior planet whose mass and orbit he hoped to derive.[19] As a preliminary he assigned one of his pupils, F. W. Flemming, to carry out reductions of the observations of Uranus. Tragically, at 28, Flemming died leaving his calculations unfinished, and they were not published until 1850. Although Bessel told Sir John Herschel in 1842 that "Uranus is not forgotten," he himself became seriously ill and was unable to make further progress. He died in March 1846.[20]

In 1842, the problem of Uranus was proposed as the subject of the prize essay of the Royal Academy of Sciences of Göttingen, with special reference to the large and still increasing errors of Alexis Bouvard's *Tables*. Eugène Bouvard had yet to present his work to the Paris Académie of Sciences, his approach being thorough and cautious, when, in June 1845, Arago put the problem in Le Verrier's way.

Eugène finally presented his still unpublished work to the Académie in September 1845. He began by noting the discrepancies in longitude in his uncle's *Tables* had by now accumulated to 2 arc minutes. Rather unexpectedly, the earlier records (from Flamsteed's of 1690 up to those made at the end of the 18th century) did not prove to be the main source of difficulty. They could, he showed, be satisfied by assuming a perfectly regular motion for Uranus (due allowance having been made, of course, for perturbations by known planets). The problem was not, as Alexis Bouvard had assumed, with the old observations but with the new ones. The discrepancies in the planet's motion, nonexistent prior to 1800, had increased to a maximum in 1822 and in that year changed their sign. Prior to that, Uranus was moving slightly ahead of its predicted position. Afterwards it was beginning to lag behind by ever greater amounts—20 arc seconds in 1830, 90 arc seconds in 1840, 120 arc seconds in 1844. By painstakingly reworking the calculations, Eugène was able to reduce the discrepancy to only 15 arc seconds, but even this was too much to be accounted for by observational errors, and he concluded in favor of his uncle's idea of a perturbing planet.[21]

For all Eugène Bouvard's long absorption in and masterful analysis of the problem, his tables were never published: He was completely co-opted by Le Verrier, who was capable of Bouvard's thoroughness but combined with some of Arago's flair. Le Verrier presented the results of his first skirmish, "Premier mémoire sur la théorie d'Uranus," to the Académie in November 1845.[22] He had by then carried out his own reworking of the motion of Uranus, which began by discrediting Alexis Bouvard. The latter had made numerous technical errors in calculations, and Le Verrier pointed them out one by one.[23] Le Verrier also took Bouvard to task for having doubted the old observers. "One should note," he wrote, "that, though the observations of Flamsteed, Bradley, Mayer, and Lemonnier are not as exact as the astronomers of our epoch, one may not with any plausibility be allowed to consider them infested with such enormous errors as those of which the present tables accuse them."[24] Fortunately Alexis Bouvard had been dead since 1843 and so was spared this biting criticism. But Le Verrier could be equally scathing toward living colleagues and pointed out the faults in some of Eugène's work at the same time. (Both Le Verrier and Eugène were up for nomination to a vacancy in the Académie. The attack on Eugène's work was presumably timed to influence the outcome. When the votes were counted, Le Verrier had resoundingly trounced his rival: 44 votes to 9, with 2 abstentions. Behind a restless intellect lurked a ruthless ambition.)

Briefly Le Verrier hoped that once all the errors made by Alexis Bouvard were corrected, the discrepancies in Uranus's motion would remedy themselves. After his laborious calculations, Le Verrier still found an outstanding error of 40 arc seconds. He had supreme confidence in the theory of gravitation. The conjecture that Uranus's enormous distance from the Sun somehow affected its authority could not be entertained and should, he declared, be used only as "a last resort." There were other possibilities—the interference of a resisting medium, for instance—but since they could not be treated mathematically he had nothing to say about them. Instead, he decided to narrow his own search to what could be defined by calculation: the proposition that the elements of Uranus used as a basis of Bouvard's *Tables* were wrong because they failed to take into account the action of an unknown planet.

Le Verrier thus came to attack the problem as Arago had suggested, by searching directly for the elements of the perturbing body needed to supply corrections to the motion of Uranus. He set for himself a problem in inverse perturbations, something whose method of solution was at the time (so far as he knew) wholly untried.

Without simplifying assumptions, the problem was dauntingly complicated. It involved the solution of an equation in no less than 13 unknowns, since the errors in Uranus's motion derive from two sources: those in the elements of Uranus itself and those produced by the perturbations of the unknown body with unknown mass and elements. Since these errors are inextricably intertwined, independent treatment is impossible.

Le Verrier assumed the orbit of the unknown planet would be noncircular, but made simplifying assumptions about the distance and the inclination. The planet must, he reasoned, be exterior to Uranus, since otherwise its effects on Saturn would be more perceptible than observation or theory allowed. Initially, he assumed a distance for the hypothetical planet of twice that of Uranus, and an orbit either negligibly or not inclined to that of the perturbed planet. Thus he came to these related, but not equivalent, questions: Given such a planet, could the discrepancies in the motion of Uranus be accounted for by its perturbing effect upon it, "and if so, exactly where is that planet? What is its mass? What are the elements of the orbit it describes?"

His simplifying assumptions collapsed the problem from an equation in 13 to one in only nine unknowns. To solve nine unknowns algebraically, nine independent equations were needed. These independent equations, or equations of condition, were formed by setting the expressions containing the elements of the perturbed and perturbing planets as defined by perturbation theory, equal to the residuals. That is, they were equal to the differences between the observed and theoretical positions of Uranus.

Le Verrier first attempted to set up equations of condition based on observations from 1747 to 1845, divided into eight equal periods. Almost immediately he saw the need to increase the time span still further, so he added Flamsteed's observation of 1690. He now sought the mass of the unknown planet by means of a laborious trial and error calculation in which, for epoch January 1, 1800, he assumed different longitudes for the planet, separated by 9° intervals along the orbit so that the whole possible range from 0° to 360° was covered.[25] All values of the mass either negative or so large as to cause a perceptible perturbation in the motion of Saturn, which had not been observed, were thrown out. This left two acceptable solutions in which the planet lay between heliocentric longitudes 108° and 162° or between 297° and 333°. Alas, his further calculations showed that neither of these solutions seemed able to satisfy the 1690 or 1747 observations. He seemingly lost the thread and concluded "it was impossible to represent the course of Uranus by means of the perturbative action of the new planet."[26]

At this point Le Verrier's investigation "ground to a halt."[27] Lost in an ungodly blizzard of arithmetic, he temporarily laid the matter aside. He could afford to do so, believing he was working alone. The fact was otherwise. The problem of Uranus was very much in the air, the attention of astronomers inescapably drawn to it by the "spirit of the age." It had seized not only the attention of the Göttingen prize committee and of the leading astronomers of France, but also that of a brilliant but obscure young Cambridge mathematician, John Couch Adams (Figure 10).

Adams had been the first to blaze the trail. He had arrived at a solution by September 1845—the time when Eugène Bouvard presented his results to the Académie and 2 months before Le Verrier finished his systematic demolition of Alexis Bouvard's *Tables*.

10. John Couch Adams (1819–1892). (Courtesy of the Mary Lea Shane Archives of the Lick Observatory.)

Adams was born in 1819 at Lidcot on Laneast Down in Cornwall. It was a bleak open moor, in sight of the sea and the Dartmoor Hills.[28] His father was a tenant farmer. At first young Adams was educated by a local schoolmaster, inauspiciously named Mr. Sleep, who according to his advertisement "challenge[d] any man in England for Calligraphy, Stenography or the Mathematics." However, Mr. Sleep was soon surpassed by his pupil, who mastered the one algebra text he had. At 12, Adams left for a better school at Devonport, run by the Rev. John Couch Grylls, a cousin of his mother's.

At Devonport, much of Adams's time was spent reading in the Mechanics Institute and teaching himself mathematics.

A noteworthy event took place in 1835, when Adams observed the return of Comet Halley and exulted to his parents, "You may conceive with what pleasure I viewed this, the first Comet I ever had a sight of, which at its visit 380 years ago threw all Europe into consternation, but which now affords the highest pleasure to astronomers for proving the accuracy of their calculations and predictions."[29]

The question of his future now arose. The logical thing was for him to go to college, but the family could ill afford the expense. At one point, Adams seriously considered setting off for Australia with Rev. Grylls and his family. However, Adams's parents insisted that regardless of the financial hardship school would cause them, he must go to Cambridge, so to Cambridge he went. In 1839 he entered St. John's College, with its "three gloomy courts," and not far from Trinity with its "loquacious clock" and

> . . . antechapel where the statue stood
> Of Newton with his prism and silent face,
> The marble index of a mind for ever
> Voyaging through strange seas of Thought,
> alone.[30]

Adams, at 20 (a year older than most of the other entering students), performed well enough on the entrance examinations to win a scholarship. Indeed, his standards were always very high, and he was never challenged for the first place in scholarship. Despite his academic success, he remained modest and unassuming. Another undergraduate at St. John's later remembered him as "rather a small man, who walked quickly, and wore a faded coat of dark green."[31]

Adams had never taken his astronomy interest seriously until June 1841. While browsing in a Trinity Street bookshop, he happened across Airy's *Report* of 1831–1832, with its mention of the problem of Uranus. A few days afterwards he jotted in his diary:

1841 July 3. Formed a design, in the beginning of this week, of investigating, as soon as possible after taking my degree, irregularities in the motion of Uranus, wh[ich] are yet unaccounted for; in order to find whether they may be attributed to the action of an undiscovered planet beyond it; and if possible thence to determine the elements of its orbit, etc. approximately, wh[ich] w[oul]d probably lead to its discovery.

In January 1843, he passed his degree brilliantly and was elected a fellow of St. John's. Only then did he begin to think seriously about the problem he had defined in his notebook. Although some, he wrote later:

> had even supposed that, at the great distance of Uranus from the sun, the law of attraction becomes different from that of the inverse square of the distance . . . the law of gravitation was too firmly established for this to be admitted till every other hypothesis had failed, and I felt convinced that in this, as in every previous instance of the kind, the discrepancies which had for a time thrown doubts on the truth of the law, would eventually afford the most striking confirmation of it.[32]

He now turned all his attention upon the idea of an exterior planet.

Like Le Verrier's, Adams's formulation of the problem involved setting up equations of condition from which to find the unknowns. Adams began with an assumption about the unknown planet's distance from Bode's law—38.4 AU. He also assumed that the orbit was circular and of low inclination.

In 1843, when Adams returned to Lidcot during the long summer vacation, he threw himself into the work of carrying out the intricate calculations. These involved working out (given his simplifying assumptions) the rest of the elements of a preliminary orbit for the unknown planet, seeing how far this served to reduce the residuals of Uranus, and repeating the process until by successive approximations he brought the theory to within the limits of error of the observations. He began only with the "modern" observations—those made since 1781. His younger brother Thomas later recalled:

> Frequently, night after night, I have sat up with him in our little parlour . . . when all else had gone to bed looking over his shoulder seeing that he copied, added, and subtracted his figures correctly to save his doing it twice over. On those occasions [our] dear Mother, who would be exhausted with her heavy work, before going to bed would

prepare the milk and bread for us for supper before retiring. This I
should warm when required and we take together. . . . Often have I
been tired, and said "It's time to go to bed, John." His reply would
be, "In a minute," and go on almost unconscious of anything but his
calculations. In his walks on those occasions on Lanseat Down, often
with me, his mind would be fully occupied in his work. I might call
his attention to some object, and get a reply, but he would again re-
lapse into his calculations.[33]

Adams's first sketch of the problem was encouraging. "The
result shewed that a good general agreement between theory and
observation might be obtained," he noted. The agreement was
least satisfactory for the years when the observations used were
fewest in number (1818–1826), and so through Rev. James Challis
(Figure 11), the overworked but affable director of the Cambridge
Observatory, he applied to Airy for additional Greenwich Observatory positions for those years. Airy was more than obliging, answering by return mail and sending not only the observations
Adams had requested but all the Greenwich observations from
Bradley's prediscovery observation of 1754 to those made in 1830.

Unfortunately, when this new data arrived from Airy, Adams
was in the midst of his usual college duties. By spring 1844, when
the Göttingen prize was announced, he again protested his duties
at college prevented him from carrying out the complete examination necessary to write a full paper on the Uranus problem.
Nevertheless, he did begin a second analysis. For the first time he
included the old observations as well as those made since 1781. He
accepted the results of Bouvard's *Tables*, but recomputed them for
the perturbations of Jupiter and Saturn to assure himself there
were no outstanding errors, but (unlike Le Verrier) found none of
any consequence, except for one in a single equation previously
pointed out by Bessel. For the modern observations, the errors
were taken exclusively from the Greenwich observations, with the
exception of one by Bessel from 1823. Although Adams retained a
distance based on Bode's law for the suspected planet, he abandoned the assumption of a circular orbit and proceeded to repeat
the laborious calculations for a new set of equations of condition,
taking into account small perturbations due to the eccentricity.

11. James Challis (1803–1882). (Courtesy of the Mary Lea Shane Archives of the Lick Observatory.)

His work on Uranus was interrupted by his calculations of an orbit for newly discovered Comet de Vico, which he published as a letter to the London *Times*.[34] He did not return to it afresh until after attending the Cambridge meeting of the British Association for the Advancement of Science in June 1845. Inspired upon "seeing for the first time some of our greatest scientific men, [Sir John] Herschel, Airy, Hamilton, Brewster, etc,"[35] he redoubled his efforts and was able to finish his new set of calculations on the "supposed new Planet" by mid-September 1845. His elements at that time were:

Mean distance, a	38.4 AU
Mean annual motion, n	$1° 30'.9$
Mean longitude at epoch, Oct. 1, 1845, ϵ	$323° 24'$
Longitude of perihelion, $\tilde{\omega}$	$315° 55'$
Eccentricity, e	0.1610
Mass; Sun=1	0.0001656

Adams turned the results over to Challis. Since his calculations included a position for the planet (its mean longitude; the position in the sky where he believed it would be found on October 1, 1845), and the Cambridge Observatory had a fine telescope at its disposal (the 11.7-inch Northumberland refractor), it might be expected someone there would immediately attempt to search for the planet. However, the overworked Challis did not take Adams's hint because as he later admitted, "he had too little confidence in the indications of theory, though perhaps as much as others might have felt in similar circumstances, and with similar engagements."[36] Of course Adams himself might equally well have mounted such a search. Apparently it never occurred to him.

Challis suggested Adams pass his results on to Airy. Unfortunately, Adams was not a good correspondent; instead of writing, he decided to deliver the results to Airy in person. After all, he was just about to leave for a visit to Cornwall. When he got to Greenwich, he found Airy was away in France. Adams was disappointed, but probably not surprised.

Adams left the letter of introduction Challis had written for him, which Airy found on his return. Airy's interest was piqued, and he wrote at once to Challis: "Would you mention to Mr. Adams that I am very much interested with the subject of his investigations, and that I should be delighted to hear of them by letter from him?"[37] At the end of his vacation, Adams again passed through Greenwich on his way from Cornwall to Cambridge. On October 21, he called twice—again unannounced; the first time Airy was out, but Adams promised to call again later that afternoon. Airy's butler never communicated this to the Astronomer Royal, who was then at dinner (he always dined promptly at half past three in the afternoon; astronomers' hours are often unusual). Thus Adams left without seeing Airy. Although Adams wrote up a brief sketch of his calculations, he departed Greenwich disappointed and with the distinct impression—completely unfounded, as it turned out—he had been snubbed.[38]

Ten years in his position as Astronomer Royal, Airy was a first-class mathematician and an able, if severe, administrator. He had been senior wrangler at Cambridge, Plumian professor and director of the Cambridge Observatory at 27, and Astronomer Royal at 34. In the latter position, he completely reorganized the Greenwich Observatory, which had been a rather dismal institution under his predecessor, the Rev. John Pond—the only Astronomer Royal ever asked to resign. As necessary prerequisites to such an undertaking, Airy possessed a rare passion for organization, a prodigious memory, and a remarkable attention to detail. He insisted on efficiency—what some considered an "inflexible routine." No doubt he was personally obsessive; a younger assistant on his staff later accused him of carrying "his love of method and order . . . to an absurd extreme," such as when he devoted an entire afternoon to labeling a number of wooden cases "empty"![39] Moreover, his love of bureaucracy "militated—was almost avowedly intended to militate—against the growth of real zeal and intelligence in the staff."[40] On the other hand, there is no denying many of the reforms he undertook were long overdue, that he brought the work of the observatory to a higher standard than it had ever known, and that he was "the most innovative and dynamic man to hold the office of Astronomer Royal since its creation in 1675."[41]

Despite an almost infinite capacity for hard work, even Airy had his limits. At the time of Adams's visit he was a much harassed man, under great physical and mental strain. Above and beyond ensuring that all the "daily observations" were taken, which reflected the Royal Observatory's mainly utilitarian function to the Royal Navy, his services as a consultant on nonastronomical governmental matters, such as determining the optimum gauge for new railways, imposed heavily upon him in that fall of 1845. Moreover, his wife was expecting. As Allan Chapman has put it, "the last thing the Astronomer Royal could have wanted was to be unexpectedly solicited by a junior Cambridge scholar with a mathematical speculation about the perturbations of Uranus."[42]

Even so, he tried to do his duty; within 2 weeks of Adams's visit, he followed up with a letter that shows he had perused Adams's calculations. "I am very much obliged by the paper . . . you left here a few days since, shewing the perturbations on the place of Uranus produced by a planet with certain assumed elements. The latter numbers are all extremely satisfactory," he wrote.[43] Apparently he surmised Adams had simply assumed all of the elements and then shown that a hypothetical planet with such elements *might* explain the errors of Uranus—an interesting but rather speculative exercise. There was, in any case, nothing to suggest to Airy that matters had reached the stage of directing the pointing of a telescope.

As Airy saw, Adams had adopted the most critical element—the planet's distance from the Sun, or the length of its radius vector—from Bode's law. Such an assumption, as Adams readily admitted, was necessary, since without the radius vector it was fruitless to try to calculate the mass, velocity, or period of the planet. And yet to Airy, as Chapman has pointed out, this was "intellectually sloppy. . . . If Newton's laws were universal, then the solar system must be held together by precise, measurable phenomen[a] susceptible to mathematical *description,* and not to a mere chance sequence of numbers, as seemed to be the basis of Bode's law. It was not that Airy in any way doubted gravitation, but that one might well come to doubt it if new planets could be discovered by what he saw as a species of unsubstantiated numerology."[44]

Indeed, Airy had long been obsessed by the radius vector of Uranus. As already noted, he found it always came out too small—this by an amount considerably greater than the distance between the Earth and the Moon. Although Adams's calculations showed his hypothetical planet would account for the error in longitude, it did not follow that the same hypothesis would automatically account for the error in the radius vector. Airy queried Adams on the matter; it was to him, he later declared, the critical question, "truly an experimentum crucis," on which he believed would turn the continuance of the law of gravitation itself. "I waited with much anxiety for Mr. Adams's answer," Airy later recalled. "Had it been in the affirmative, I should at once have exerted all the influence which I might possess, either directly or indirectly through my friend Professor Challis, to procure the publication of Mr. Adams's theory."[45] But Adams never replied; perhaps, as Challis later surmised, from sheer procrastination, more likely because he thought there was no use in doing so, that Airy was simply trying to put him off. Adams later told Challis he regarded Airy's question as "trivial," a mere detail.[46] But the devil is in the details. His failure to follow up is one fault that cannot be laid at Airy's door.

> The fault, dear Brutus, is not in our stars,
> But in ourselves, that we are underlings.
> —William Shakespeare
> *Julius Caesar,* Act I, Scene ii

Whatever his motivations (or lack thereof), Adams's failure to reply effectively brought an end to the correspondence with the Astronomer Royal for nearly a year. They actually met twice during that time: once in December 1845, when Adams was with Challis, but Airy later forgot where, and in July 1846 by chance, and for no more than 2 minutes, on St. John's Bridge. Adams seems to have made virtually no impression on Airy at all. Although Adams continued to work at refining his calculations, he did so in the isolation of his Cambridge redoubt; his results went unpublished, and the possibility of keeping the discovery of the new planet entirely in English hands, which had briefly gleamed in the fall of 1845, was lost forever.

The new year began. It was 1846, the year of Francis Parkman and the Oregon Trail. Le Verrier meanwhile was forging his own Oregon Trail into the realms of abstruse mathematical investigations. He succeeded in breaking the impasse that had stalled him for several months. By neglecting two minor terms and repeating his calculations, he revised his second solution for the unknown planet and satisfied himself it was probably located at mean longitude between 243° and 252° on January 1, 1800, with the latter the most likely value. This would lead to a longitude of 325° for January 1, 1847. Le Verrier's position was within little more than 1° of that already reached, but still unpublished, by Adams.

Although late coming, Le Verrier was first to publish. His solution appeared (though without a full set of elements specifying the planet's orbit) in his second memoir, "Recherches sur les mouvements d'Uranus" (Researches on the movements of Uranus), in the *Comptes rendus* for June 1, 1846.[47] Airy received the paper on June 23 or 24. After the singular want of sympathy he had felt toward the work of Adams, his attention was suddenly seized. Le Verrier, after all, was a strong personality, his reputation as a mathematician already established and his abilities known firsthand to Airy himself. The agreement of the two independently calculated positions was nothing less than astounding. "I cannot sufficiently express," Airy wrote later, "the feeling of delight and satisfaction which I received" from Le Verrier's calculation:

> The place which it assigned was the same, to one degree, as that given by Mr. Adams's calculations, which I had perused seven months earlier. To this time I had considered that there was still room for doubt of the accuracy of Mr. Adams's investigations; for I think that the results of algebraic and numerical computations, so long and so complicated as those of an inverse problem of perturbations, are liable to many risks of error in the details of the process . . . . But now I felt no doubt of the accuracy of both calculations, as applied to the perturbation in longitude.[48]

Airy wrote a day or two later to William Whewell, master of Trinity College, Cambridge, who was vacationing at Lowestoft, on the remarkable agreement of these results. Airy added: "if I were a rich man or had an unemployed staff I would immediately take measures for the strict examination of that part of the heavens containing the position of the postulated planet."[49] On June 29, he declared at a meeting of the Board of Visitors of the Royal Observatory, whose members included Challis, Sir John Herschel, and Capt. W. H. Smyth, of the "extreme probability of now discovering a new planet in a very short time, provided the powers of one observatory could be directed to the search for it."[50]

By now, Airy had queried Le Verrier about the error in the radius vector—the problem about which Adams had failed to reply. Le Verrier did reply, immediately and decisively. In effect, he put the blame on Bouvard: "One must realize that the orbit of Uranus has been calculated by M. Bouvard using some positions of the planet that do not fit with an elliptical orbit, since he had no way of taking into account the perturbations due to the unknown planet. This circumstance has necessarily rendered the elements of his orbit incorrect, and it is the errors of his eccentricity and longitude of the perihelion which account for the present error in the radius vector of Uranus. . . . Excuse me, Sir, for insisting on this point."[51] Airy's scruples were at last satisfied. "It is impossible," he later explained, "to read this letter without being struck with its clearness of explanation, with the writer's extraordinary command . . . of the physical theories of perturbation, and with his perception that his theory *ought* to explain all the phenomena, and his firm belief that it *had* done so. I had now no longer any doubt upon the reality and general exactness of the prediction of the planet's place."[52]

Airy's actions or inactions, with which one can sympathize up to this point, now become somewhat inscrutable. Le Verrier, in responding to the query about the radius vector, had added, "If, as I hope, you have sufficient confidence of my work in order to look for this planet in the sky, I would, Sir, speedily send you its exact position, as soon as I have obtained it." Airy however pulled an Adams; he did not respond to Le Verrier's letter. He later justified his silence by saying there was no point in troubling Le Verrier for a more accurate position given that he, Airy, was about to leave for the continent. This is certainly disingenuous; in fact, knowing of Adams's result, he had no need for further numbers. And yet he must have realized in responding to Le Verrier's letter he could hardly avoid mentioning Adams's investigation. Faced with these alternatives, he chose silence. It was this omission he had the most difficulty explaining away later.[53]

He certainly grasped the opportunity that offered itself. Within days of seeing Le Verrier's paper, he quickly drew up a search plan for the planet such as Le Verrier had called for. He did not, however, delegate it to the Royal Observatory at Greenwich, as might be expected, but instead to the observatory at Cambridge. He followed up his remarks to the Board of Visitors by writing Challis on July 9: "You know that I attach importance to the examination of that part of the heavens in which there is a possible shadow of reason for suspecting the existence of a planet exterior to Uranus. I have thought about the way of making such examination, but I am convinced that (for various reasons, of declination, latitude of place, feebleness of light, and regularity of superintendence) there is no prospect whatever of its being made with any chance of success, except with the Northumberland telescope." At 11.75 inches, the Northumberland telescope had a greater light-grasp than the 6.75-inch Sheepshanks telescope of the Greenwich Observatory. Realizing Challis was always absurdly overworked and might not be able to pursue this project with dispatch, Airy offered to put at his disposal an assistant from the Greenwich Observatory staff. He concluded: "You will readily perceive that all this is in a most unformed state at present, and that I am asking these questions, almost at a venture, in the hope of rescuing the matter from a state which is, without the assistance that you and your instruments can give, almost desperate. . . . The time for the said examination is drawing near."[54]

Airy's puzzling indirectness ("a possible shadow of reason for suspecting the existence of a planet, . . . all this is in a most unformed state") was hardly calculated to evoke an enthusiastic response from Challis, and betrays perhaps an uneasiness of mind. In any case, Challis was away when Airy's letter arrived. Airy, meanwhile, had had second thoughts. Perhaps realizing the Cambridge astronomer might prove insusceptible to subtle suggestion, he put more of a point on the matter in another letter drafted 4 days later. "I only add at present," he wrote, "that, in my opinion, the importance of the inquiry exceeds that of any current work, which is of such a nature as not to be totally lost by delay."[55] Airy called upon Challis to examine a swath of sky centered on the theoretical position of the planet some 30° long and 10° wide. (He did not, however, insist on what had been an essential part of his system at Greenwich—that reduction of the observations proceed *pari passim* with the observations themselves, a critical omission.)

Before Airy left England for the continent (August 10), Challis informed him in a letter dated July 18 of his intent to personally undertake the search—thus declining Airy's offer of an assistant—and had begun plotting stars in the search zone on July 29. Airy probably felt he had done all he could do to help in discovering the new planet, a process he expected to be long and laborious. At the outset, he estimated Challis might have to spend 300 hours scanning stars in the search zone.

Though Adams had indicated the planet would be no fainter than 9th magnitude, Challis decided to include stars to 10th and 11th magnitude. In the area of sky encompassed by his search, there are more than 3,000 stars 11th magnitude or brighter. His plan was to plot each star in the search zone three times, and then to compare their positions. The method was essentially the same as that used the previous December by Hencke to discover the fifth asteroid Astraea. A planetary body would reveal itself by its motion.

During August Challis plodded, slowly and cautiously, with the Northumberland, and on September 2 wrote to update Airy, who was then at Wiesbaden, in Germany: "I have lost no opportunity of searching for the planet; and, the nights having been generally pretty good, I have taken a considerable number of observations; but I get over the ground very slowly, thinking it right to include all stars to 10–11th magnitude; and I find, that to scrutinize, thoroughly, in this way the proposed portion of the heavens, will require many more observations than I can take this year."[56] He put off comparing the observations until later; "partly," he explained, because "he had too little confidence in the indications of theory."[57] He later found he had actually recorded the planet twice, on August 4 and 12, but had failed to recognize it for what it was.

Events now began to move quickly, and in the end overtook the plodding Challis. On August 31, Le Verrier published a paper in which he finally gave orbital elements for the hypothetical planet, noting opposition occurred on or about August 19. He located its probable position some 5° east of the third-magnitude star δ Capricorni, adding that it ought to show a disk 3.3″ in diameter. "This is an important point," he emphasized. "If . . . the planet's disk has a large enough diameter to preclude its being confused with the stars, if one can substitute for the rigorous determination of the positions of all luminous points, a simple inspection of their physical appearance, then the search will move along rapidly."[58]

As Le Verrier's paper did not arrive in England until the latter part of September, it had no immediate influence there. Meanwhile, Adams was stirring again. Unaware Airy was out of the country and apparently still unaware he had a rival, he requested more observations of Uranus from Greenwich (in Airy's absence, his assistant, Rev. Robert Main, sent the latest observations for 1844 and 1845). Adams also included a revised set of orbital elements. It is of interest to compare the latest orbits calculated by Adams and Le Verrier side by side:

	Adams (Sept., 1846)	Le Verrier
Mean distance, a	37.25	36.15
Mean longitude, ϵ 1847.0	329° 57′	326° 32′
Longitude of perihelion, $\bar{\omega}$	299° 11′	284° 45′
Eccentricity, e	0.12062	0.10761
Mass, Sun=1	0.000150	0.0001073

Adams proposed to give a brief account of his investigation of Uranus at the meeting of the British Association of the Advancement of Science, at Southampton, but his ill luck continued; he arrived, paper in hand, on September 15, exactly one day too late; the section on Mathematical and Physical Science had already closed. Had he arrived at Southampton September 10 he would have heard Sir John Herschel announce: "We see it [the planet] as Columbus saw America from the far shores of Spain. Its movements have been felt, trembling along the far-reaching line of our analysis with a certainty hardly inferior to ocular demonstration."[59] But Herschel too had failed to make public mention of Adams's name.

Airy had supposed the unknown planet would be inaccessible in ordinary telescopes, and thus delegated it to the Northumberland. Perhaps the Northumberland did stand the best chance, but there seems to have been another influence at work—what Robert W. Smith has called the "Cambridge Network." The members of the Board of Visitors at Greenwich in June 1846 were mainly Cambridge men, and it seems undeniable that Airy and the rest were motivated by considerations of loyalty to the old school tie in their hopes of securing for Cambridge a "double triumph: Adams's theoretical discovery and Challis's optical discovery."[60] This is only natural and goes far toward explaining Airy's failure, otherwise puzzling, to respond to Le Verrier's suggestion for a search, and indeed the whole attitude of secrecy surrounding the Cambridge project during the lost summer of 1846; what John Russell Hind would refer to as "the inexcusable secrecy observed by all those acquainted with Adams's results."[61]

By mid-September 1846, 24-year-old Hind, in charge of the 7-inch refractor at a private observatory at Regent's Park, London, became a belated entrant into the quest for the planet, having taken a hint from Challis. As diligent as he eventually proved himself to be—in the 1850s, he became well-known as a discoverer of asteroids—he would presumably have found the planet had he been informed earlier. Hind was in communication not only with Challis but also with Hervé Faye at the Paris Observatory, from whom he learned ahead of anyone else in England of Le Verrier's suggestion to scout out the planet's disk instead of sifting through star positions. He attempted to pass the idea along to Challis, but the latter did not immediately seize on its importance.[62] Faye meanwhile hinted to Hind that a brief search for Le Verrier's planet had already taken place at the Paris Observatory, but the observers there—not armed, as Airy had been, with the confidence of a double prediction by two independent calculators—had abandoned it by August 12. Now Faye was beginning to meditate on a renewed search of his own. Also, Sir John Herschel had put something about the planet to his friend William Rutter Dawes, who in turn mentioned it to William Lassell, the wealthy brewer and amateur astronomer at Liverpool, who had just set up a splendid 24-inch reflector. Neither Dawes nor Lassell mounted a search for the planet, possibly in part because Herschel confused the date Le Verrier cited for computational purposes (January 1, 1847) as that when the planet would be best seen. Thus Herschel suggested "it is not generally supposed that the stranger would show itself till about Christmas."[63] Moreover, Lassell was "confined to a sofa with a sprained ankle" when Dawes's letter arrived. In the interim, the letter was "misplaced through the carelessness of a maid-servant."[64] By then, Lassell did not have time to follow up the prospect that had briefly glimmered before his eye. Nevertheless, in spite of miscues and missed opportunities, the net around the planet was tightening.

Le Verrier, of course, was completely unaware of developments in England. As far as he knew, Airy had done nothing either to pursue his hint of the previous June. Nor had his suggestion that astronomers search for a small planetary disk among the stars stirred obvious interest. On September 18, Le Verrier wrote to Johann Gottfried Galle (Figure 12), then an obscure astronomer at the Royal Observatory in Berlin. A year earlier, Galle had sent Le Verrier his doctoral dissertation, which concerned observations made by 17th-century Danish astronomer Olaus Roemer. Belatedly, Le Verrier wrote to acknowledge it. Among other things, he queried Galle about Roemer's Mercury observations, but then came quickly to his point:

> Right now I would like to find a persistent observer, who would be willing to devote some time to an examination of a part of the sky in which there may be a planet to discover. . . . You will see, Sir, that I demonstrate that it is impossible to satisfy the observations of Uranus without introducing the action of a new Planet, thus far unknown; and, remarkably, there is only one single position in the ecliptic where this perturbing Planet can be located. . . . The actual position of this body shows that we are now, and will be for several months, in a favorable situation for the discovery.[65]

Galle indeed proved to be his man. He received Le Verrier's letter on September 23, and at once sought permission from the observatory's director, Johann Encke, to carry out the search. Encke was skeptical but nonetheless acquiesced: "Let us oblige the gentleman in Paris." A young student astronomer, Heinrich Ludwig d'Arrest, begged to be included, and joined Galle as a volunteer observer. That night, they opened the dome to reveal the observatory's main instrument, a 9-inch Fraunhofer refractor aimed at the

12. Johann Gottfried Galle (1812–1910). (Courtesy of the Astronomical Society of the Pacific.)

spot assigned by Le Verrier. Recalculated for geocentric coordinates, its position was at right ascension 21 h, 46 min, declination −13° 24′, very close to the position occupied by another planet, Saturn.

The question arose: What maps were available? At first they could think of none but "Harding's very insufficient Atlas." D'Arrest then suggested "it might be worth looking among the Berliner Akademische Sternkarten to see whether Hora XXI was among

those already finished. On looking among a pile of maps in
Encke's hall [Vorzimmer], Dr. Bremiker's map of Hora XXI [al-
ready engraved and printed at the beginning of 1846 but not yet
distributed] was soon found." As d'Arrest later recalled, "We then
went back to the dome, where there was a kind of desk, at which
I placed myself with the map, while Galle, looking through the refractor, described the configurations of the stars he saw. I followed
them on the map one by one, until he [Galle] said: and then there
is a star of the 8th magnitude in such and such a position, whereupon I immediately exclaimed: that star is not on the map!"[66]
Neptune, the eighth planet, had at last been found, right ascension
21 h, 53 min, 25.84 sec (Figure 13).

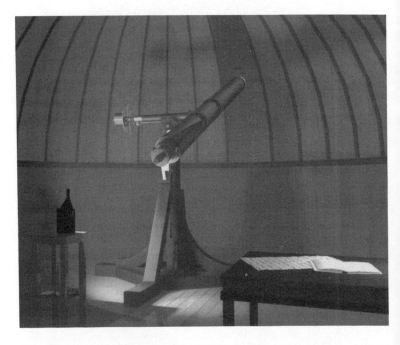

13. Midnight September 23, 1846, Berlin Observatory. Artist's
impression of the moment following the discovery of Neptune. The observers have gone to report the find to Encke, and the telescope is left
alone in quiet contemplation of the new world. (Courtesy Julian Baum.)

eight

Triumph and Controversy

D'Arrest's excited report brought Encke to the dome. The three astronomers tracked the planet until it set about 2:30 A.M. The following night they returned to this small theater of immense happenings to confirm the observation. Turning the telescope to the same general area of the sky, Galle placed his eye to the finder:

> Four stars of the eighth magnitude occupied its field. One of them was brought into the field of the large telescope and critically examined by my assistant and rejected. A second star was in like manner examined and rejected. A third star rather smaller and whiter than either of the others was brought to the center of the field of the great telescope, when my assistant exclaimed: "There it is! there is the planet! with a disk as round, bright, and beautiful as that of Jupiter." Galle himself, taking his turn at the eyepiece, exclaimed, "My God in heaven, this is a big fellow!" [1]

Galle and Encke attempted to measure the planetary disk. Encke's final result—3.2 arc seconds—was almost identical with the 3.3 arc seconds Le Verrier had predicted.[2] This indicated a giant world, with a diameter some four times that of Earth's. Careful inspection revealed the object's unmistakable motion of about 3 arc seconds per hour, retrograde. Its position in the sky was only

55 arc minutes—less than two full Moon diameters from that indicated by Le Verrier's calculations.

The last shadow of doubt had been removed. On September 25, Galle wrote to Le Verrier: "Monsieur, the planet of which you indicated the position really exists" [réellement existe]. He also mentioned a possible name for the new planet—*Janus*. Le Verrier wrote in reply thanking Galle for "the alacrity with which you applied my instructions to your observations. . . . We are thereby, thanks to you, definitely in possession of a new world. . . . Allow me to hope that we will continue to carry on a correspondence which begins under such auspicious circumstances."[3] In a postscript he mentioned the Bureau des Longitudes had decided on the name Neptune for the planet. "The name Janus," he added, "would imply that this is the last planet of the solar system, which we have no reason at all to believe." Apparently Le Verrier had mistaken Janus for Terminus, the god of limits. There is little doubt the choice of the name Neptune was Le Verrier's own; there is no evidence the question of naming the planet had yet been taken up by the Bureau des Longitudes, and it seems only too likely that Le Verrier thought, as Grosser has suggested, "his glory would be diminished if Galle's suggested name for the planet was adopted. A bogus official sanction was one way to discourage Galle at the outset."[4]

Meanwhile, Encke dispatched an official report of the discovery to Heinrich Christian Schumacher, editor of the German journal *Astronomische Nachrichten*, at Altona,[5] and by letter also announced the discovery to Peter Andreas Hansen, at Gotha, with whom Airy was then staying. This arrived at Gotha on September 29, the date Airy first learned of the discovery.

In the next morning's post, the news reached England. Hind learned it from a German correspondant, F. F. E. Brünnow. He was able to observe the planet from London that night. Meanwhile, Challis had just received Le Verrier's memoir of August 31, in which the possibility of locating the planet by means of its disk had been mentioned. As noted earlier, he had uncognizantly recorded it on August 4 and 12. However, as he later recalled:

> I did not again meet with the planet till September 29, on which day
> I saw for the first time the results of M. Le Verrier's last investiga-

tions. By these I was induced to return again to the theoretical position of the planet, and to endeavour to detect it by the appearance of a disk. In fact on the night of September 29, out of a very large number of stars whose approximate places I recorded, I fixed upon one which appeared to me to have a disk.[6]

A few additional details about this incident were later provided by the Rev. William Towler Kingsley, tutor of Sidney Sussex College. He was dining with Challis at Trinity College that evening and suggested it might be worthwhile to examine the suspect with higher magnification. Challis agreed and, since the sky was clear, proposed they should go to the observatory immediately after dinner. However, Mrs. Challis insisted on serving them tea. While they delayed, the sky clouded up. The following night, September 30, the Moon was in the way. On October 1, Challis read of the great events that had taken place in Berlin in a letter Hind had written to the London *Times*. Only then did he check his observations begun on July 29, to realize the planet had been within his grasp—twice in the first 4 days of observing.

Between October 1, when he had written to Galle, and October 5, Le Verrier had changed his mind about the name for the planet; instead of Neptune, he decided it would be fitting to name the planet after himself. He beseeched Arago, "as a friend and a countryman," to adopt the name Le Verrier. Arago obliged, and in an attempt to provide a cover of modesty became the main promoter of the eponymous designation. Arago informed the Paris Académie des Sciences he had "received from M. Le Verrier a most flattering invitation, the privilege of naming the new planet. . . . I have made up my mind to call it by the name of the man who so learnedly discovered it, to name it *Le Verrier*."[7] He later attempted a passionate defense of his decision:

Comets are named after the astronomers who discovered them, or who traced out their orbits; should the same honor be refused to the discoverers of planets! . . . Is it right that we should have comets named for Gambart or Biela, de Vico, Faye, and others, and that the man who, by an admirable and unprecedented method, has demonstrated the existence of a new planet, specified its place and size, should not be written in the heavens!!! No, no! It offends reason and the most common principles of justice.[8]

For consistency, Arago also resolved to call the seventh planet Herschel instead of Uranus.

Le Verrier coyly feigned surprise at Arago's choice: "I requested my illustrious friend M. Arago to undertake the duty of naming the planet," he wrote to Encke. "I was a little surprised by the choice he had made in the intimate circle of the Academy." And yet when Le Verrier published his combined memoir on Uranus a few months later, he retitled it, "Recherches sur le mouvement de la planète Herschel (dite Uranus)"; undoubtedly, as W. G. Hoyt has surmised, an attempt "to give this name both astronomical and political legitimacy."[9] Although it was too late to change references to the planet Uranus in the text, Le Verrier added the following note: "In my future researches, I shall consider it my strict duty to eliminate the name Uranus completely, and to call the planet only by the name *Herschel*."

Indeed, the eighth planet was commonly referred to, for a time, as "Le Verrier's planet," although the leading astronomers, including Encke and Karl Friedrich Gauss, favored Le Verrier's first choice, Neptune. So did the English astronomer W. H. Smyth, who wrote with unassailable logic to Airy: "Mythology is neutral ground. Herschel is a good name enough. Le Verrier somehow or other suggests a Fabriquant and is therefore not so good. But just think how awkward it would be if the next planet should be discovered by a German, by a Bugge, a Funk, or your hirsute friend Boguslawski."[10] Eventually, the name Neptune prevailed. Nevertheless, Le Verrier received acclaim enough. He was created an officer of the Legion of Honor by King Louis Philippe and from the King of Denmark received the title of Commander of the Royal Order of Dannebroga. The Academy of St. Petersburg promised him the very first vacancy in their body, and the Royal Society of Göttingen named him Foreign Associate.

Adams's role, meanwhile, had remained completely unknown. The first mention of it in print was by Sir John Herschel. Referring to his early September comments to the British Association for the Advancement of Science, he penned a letter on October 1 to the London *Athenaeum*:

The remarkable calculations of M. Le Verrier—which have pointed out, as now appears, nearly the true situation of the new planet, by resolving the inverse problem of the perturbations—if uncorroborated by repetition of the numerical calculations by another hand, or by independent investigation from another quarter, would hardly justify so strong an assurance as that conveyed by my expression above alluded to. But it was known to me, at that time, (I will take the liberty to cite the Astronomer Royal as my authority) that a similar investigation had been independently entered into, and a conclusion as to the situation of the new planet very nearly coincident with M. Le Verrier's arrived at (in entire ignorance of his conclusions), by a young Cambridge mathematician, Mr. Adams;—who will, I hope, pardon this mention of his name.[11]

That same day, Herschel also wrote to William Lassell about the new planet. Lassell, as mentioned earlier, had been informed too late to be seriously in the running among searchers of the planet. Herschel may have been feeling somewhat guilty about this, since he had been in correspondence with Lassell about the satellites of Saturn and would have had a good chance of recognizing the unknown planet by its disk with his fine equatorially mounted 24-inch reflector, if only he had known where to look. Clearly Herschel was reeling with the news the discovery had been made in Berlin, but hoping something might yet be salvaged for England, urged the brewer–astronomer to look out for "satellites with all possible expedition!!"[12]

Lassell first observed "Le Verrier's planet" with the 24-inch reflector on October 2, and the next night suspected it was surrounded by a ring. Within another week, he had also glimpsed a "star" close by the planet, which he supposed was a satellite. Poor weather and the planet's declining altitude prevented verification until the next July. (The ring, apparently confirmed by Challis and others, was eventually shown to be illusory—it has nothing to do with the actual rings of the planet, which are far too wispy and faint to have been seen in Lassell's telescope.)[13]

Herschel's letter to the *Athenaeum* attracted little immediate attention. But the information about Adams's calculations, so long buried in secrecy, was beginning to leak out. Airy, on his return to

Greenwich from Germany, had written to Challis on October 14, "Heartily do I wish that you had picked up the planet. . . . But these misses are sometimes unavoidable."[14] He was by then writing up an account of the English researches, hitherto unknown to the world. He asked Challis for permission to publish their previous correspondence on the subject, and made the same request of "Rev. W. J. Adams." (In his reply, J. C. Adams wrote, "I may mention that I am not yet in Orders." Indeed, he never did take them.)

For the first time since the previous June, Airy communicated with Le Verrier, tendering his "sincere congratulations on the successful termination to your vast and skilfully directed labours." He added rather sheepishly:

> I do not know whether you are aware that collateral researches had been going on in England, and that they had led to precisely the same result as yours. I think it probable that I shall be called on to give an account of these. If in this I shall give praise to others, I beg that you will not consider it as at all interfering with my acknowledgment of your claims. You are to be recognised beyond doubt as the real predicter of the planet's place. I may add these English investigations, as I believe, were not quite so extensive as yours. They were known to me earlier than yours.[15]

Le Verrier reacted with shock and outrage. Referring to Herschel's letter, he complained to Airy it was "very bad and unjust toward me." He was hurt by the latter's comment that he should not have felt justified in expressing himself so confidently at Southampton if his results had not been independently corroborated by Adams's work. With regard to Adams, he asked Airy, "Why should Mr. Adams have maintained silence for so many months?" He appealed to Airy to defend him and cited documents proving that as late as September 28 and 29 Challis had indicated he was searching for the planet "from my indications." Airy desperately attempted to smooth the waters: "I am confident that you will find that no real injustice is done to you," he told Le Verrier, "and I hope that you will receive this expression the more readily from me, because I have not hesitated to express to others, as well as to yourself, very strong feeling upon the extraordinary merit of your proceedings in the matter.[16]

But it was too late. The next day, October 17, Challis's first account of his involvement was printed in the *Athenaeum*. It became generally known that he and Airy had been in possession of Adams's calculations, giving the position of the planet as early as October 1845, and that Challis had actually been searching for the planet at Cambridge when it was found at Berlin. Moreover, Challis even went so far as to propose a name for the planet—Oceanus— something that, under the circumstances, could only be regarded as presumptuous (although Arago, strangely given the violence with which he had argued for the name "Le Verrier," agreed that this name would probably win the assent of astronomers!).[17]

Now the storm finally broke. In England, Challis and Airy were savagely attacked. "Oh! curse their narcotic Souls!" exploded Adam Sedgwick, professor of geology at Trinity College, Cambridge, and longtime friend of Airy's, upon realizing the planet had through Airy and Challis's blundering missed being born a "Cambridge man."[18] (Even Sedgwick had to admit, however, that Adams himself deserved a share of the blame, since he had "acted like a bashful boy rather than like a man who had made a great discovery.") An undergraduate at Trinity mooted some of the local gossip in a letter to his father, October 31:

> There was some interesting conversation about the new Planet. . . . It appears that Mr. Adams of St. John's had made his calculations in the spring and sent them to Greenwich to Airy, the Astronomer Royal; but he paid no attention to them, and to his neglect Sedgwick attributed the loss of the discovery. He mentioned that in the summer [someone] had seen Mr. "Nep" from the Observatory, but did not recognise him as the planet they were looking for.[19]

In France, the reaction was even more extreme. The controversy over Neptune was played out not only in scientific circles but also against the tense political backdrop of Anglo–French rivalry. In 1840, the two countries had clashed over the Near Eastern crisis involving the pasha of Egypt's ambitions in Syria, then part of the Ottoman empire. France had wanted to help the pasha, but Louis-Philippe's minister to London, François Guizot, had had to back off and concede the reality, which was that it could not exert its will in a world now dominated by Britain. Just before the discovery of

Neptune, Louis-Philippe had once again tangled with Britain over the marriage of the Duke of Cadiz and his youngest son to Spanish princesses. One French journal, *L'Univers,* even suggested the English were attempting "to make us pay the wedding costs of the Spanish marriages with M. Le Verrier and his planet."[20] This was the mood when Arago, on October 19, took up the cudgel for Le Verrier before the Académie, in a meeting that was so heated that it was asked in one of the Paris journals: "Are we in the Academy of Sciences or in the Chamber of Deputies? Is it a question of the existence of M. Guizot's Ministry or M. Le Verrier's planet?"[21] On summarizing and demolishing line by line Herschel and Challis's documents, Arago objected with unusual vehemence to the fact that nowhere had they referred to a single published work by Adams; this circumstance, in his view, "was sufficient to put an end to the debate. The only rational manner of writing the history of the sciences is to rely on publications having a definite date; otherwise all is confusion and obscurity. Mr. Adams has not published a single line about his researches. . . . He does not, therefore, have the least right to figure in the history of the discovery of the new planet." He concluded: "In the eyes of every impartial man, this discovery will remain one of the most magnificent triumphs of astronomical theory, one of the glories of the Academy, and one of our country's noblest titles to the gratitude and admiration of posterity."[22]

The following month, the French journal *L'Illustration* ran a series of unpleasant caricatures of Adams, while in England, Challis and Airy were summoned to make public reports to the Royal Astronomical Society. Challis fumbled in his attempts to explain why he had not searched for the planet when he had first received Adams's calculations and why he had not examined his data earlier; he could only plead "lack of faith" in both cases. Airy's "Account of some circumstances historically connected with the discovery of the Planet exterior to Uranus," was presented in November to an "unusually numerous attendance of Fellows" of the Royal Astronomical Society. Although written from a rather self-serving point of view, it at least set forth the main facts of the matter. Airy, not surprisingly, was savagely attacked, although it must be admitted he held up admirably—in his own view, he had done his duty; no man could do more. He has generally been viewed as the main scapegoat in the Neptune affair. Interestingly, Adams continued to regard him with awe. Adams's family placed most of the blame on Challis for not conducting a more thorough search—they could never forgive the episode of the tea![23]

An interesting perspective on the whole affair was presented by the 19th-century American astronomer Elias Loomis. According to Loomis, Challis was not to blame. The prediction had been much more accurate than anyone had had a right to expect. Even Le Verrier suggested that though the most probable longitude was 325°, the planet might well lie anywhere between 321° and 335°; moreover, "if the planet should not be discovered within these limits, then we must extend our search beyond them." "Professor Challis," wrote Loomis, "proceeded like a sagacious as well as brave general. He contemplated a long siege—yet his plan rendered ultimate success almost certain. Dr. Galle took the citadel by storm—yet he had no reason to expect so easy a conquest. His success must have astonished himself as much as it did the world":

> Let us then be candid, and claim for astronomy no more than is reasonably due. When in 1846 Le Verrier announced the existence of a planet hitherto unseen, when he assigned its exact position in the heavens, and declared that it shone like a star of the eighth magnitude, and with a perceptible disc, not an astronomer in France, and scarce an astronomer in Europe, had sufficient faith in the prediction to prompt him to point his telescope to the heavens. But when it was announced that the planet had been seen at Berlin; that it was found within one degree of the computed place; that it was indeed a star of the eighth magnitude, and had a sensible disc, then the enthusiasm not merely of the public generally, but of astronomers also, was even more wonderful than their former apathy.[24]

Certainly good luck was entirely on Galle's side, and ill luck entirely on Challis's. On such chances and mischances turn great events.

Airy, too, has had his champions, most notably Allan Chapman, who pointed out "had Adams placed his work in the public domain . . . even by means of a letter to the *Times* conveying the planet's predicted place—as he had done in his investigation into the orbit of de Vico's comet in a letter to that newspaper of 15

October 1844—then Airy might well have acted differently. Such a letter would have made it a prospective piece of human knowledge which its author had troubled to lift out of his private researches and place in the public realm." There is certainly truth to this, and also to Chapman's conclusion, "it is futile to speculate what might have happened had certain persons acted differently."[25]

Eventually, the priority dispute died down. Sir John Herschel, whose letter had started the international cause célebrè, was by the end of 1846 recoiling from the attacks of Arago and other French scientists who "fly at one like wild cats." Herschel confided to Rev. Richard Sheepshanks that he was being made ill by the whole matter.[26] But he agreed, in the end, that the French deserved much of the credit: "Though *Neptune* ought to have been born an Englishman and a Cambridge man every inch of him—*Dis aliter visum* now will never make 'An English Discovery' of it do what you will."[27] The French in any case had largely created the mathematical tools to calculate planetary perturbations. "It's all French," he acknowledged in late 1846, ". . . Clairaut, D'Alembert, Laplace, Lagrange and more recently Poisson and Pontecoulant for the analysis and Bouvard for the tables which though not *quite* correct were yet correct enough to raise the hue and cry. The new planet is as much Laplace's as it is either Le Verrier's or Adams's. . . . Who made one and all of the formulae by which *both* have grappled the planet."[28] Airy too, in his "Account," had pointed to a resolution of the priority problem by recalling the work of the predecessors of Adams and Le Verrier; he suggested that the hypothesis had long been in the air, and was due not to any one man so much as to the vast working "spirit of the age":

> . . . The discovery of this new planet is the effect of a movement of the age. It is shewn, not merely by the circumstance that different mathematicians have simultaneously but independently been carrying on the same investigations, and that different astronomers, acting without concert, have at the same time been looking for the planet in the same part of the heavens; but also by the circumstance that the minds of these philosophers, and of the persons about them, had long been influenced by the knowledge of what had been done by others, and of what had yet been left untried. . . . I do not consider this as detracting in the smallest degree from the merits of the persons who have been actually engaged in these investigations.

Certainly, had the planet not turned up when it did, it would inevitably have done so soon afterwards, when Challis got around to comparing his observations.

It is interesting that Adams always gave Le Verrier full credit for the discovery, although he claimed the credit of prior independent conjecture. Since then the consensus of the astronomical world has been to divide credit equally between them. As neither Adams nor Le Verrier had participated in the priority dispute, there was no bitterness between them personally; indeed they actually met in June 1847 at a British Association meeting at Oxford, and parted as friends. John Herschel remarked, "there is not, nor henceforth ever can be, the slightest rivalry on the subject between these two illustrious men—as they have met as brothers and as such will, I trust, ever regard each other—we have made, we could make, no distinction between them on this occasion."[29]

The actual place of Neptune was only 55 arc minutes from Le Verrier's theoretical position and 2½° from Adams's (the lesser accuracy of the latter attributable in part, apparently, to Adams's misconceptions about the radius vector). To Encke, in the first blush of the ocular demonstration of the planet's existence, this agreement between prediction and observation appeared as "the most outstanding conceivable proof of the validity of universal gravitation," as he wrote in congratulating Le Verrier. "I believe that these few words sum up all that the ambition of a scientist can wish for. It would be superfluous to add anything more."[30] Airy too remarked on the astounding accuracy of the theoretical prediction: "I cannot attempt to convey . . . the impression which was made on me by the author's undoubting confidence, . . . by the firmness with which he proclaimed to observing astronomers, 'Look into the place which I have indicated, and you will see the planet well.' "[31]

A later historian of 19th-century astronomy, Agnes Clerke, after dismissing the priority debate, wrote in similar vein:

Personal questions . . . vanish in the magnitude of the event they relate to. By it the last lingering doubts as to the absolute exactness of the Newtonian Law were dissipated. Recondite analytical methods received a confirmation brilliant and intelligible even to the minds of the vulgar, and emerged from the patient solitude of the study to enjoy an hour of clamorous triumph. For ever invisible to the unaided eye of man, a sister-globe to our earth was shown to circulate, in perpetual frozen exile, at thirty times its distance from the sun.[32]

The same sentiment is captured in Norwood Russell Hanson's arresting phrase, the "Zenith of Newtonian Mechanics."[33] The discovery was above all a supreme triumph for the methods of perturbation theory and celestial mechanics itself, which had been carried thereby "into the brightest heaven of scientific achievement."[34]

To the lay public, the discovery of Neptune seemed nothing less than miraculous, an impressive demonstration, in an age when the forces of obscurantism were still formidable, of the perfection of science. "The sagacity of Le Verrier was felt to be almost superhuman. Language could hardly be found strong enough to express the general admiration."[35] It was the discovery at a desk of a planet never seen. Camille Flammarion expressed in his incomparable way the reason the prediction and discovery gripped the imagination as it did in 1846: "This scientist, this genius," he said of Le Verrier, "had discovered a star with the tip of his pen, without other instrument than the strength of his calculations alone."[36] It seemed uncanny, had all the air about it of the magician's sleight-of-hand—a substantial planet made to materialize out of the hieroglyphs crowded onto a sheet of paper!

The triumph was not, as is sometimes suggested, a salvaging of a Newtonian law shadowed by doubts. There were few doubts about its validity then, though as we shall see, there would be significant doubts later. The discovery of Neptune resolved a crisis not in gravitational theory but in an astronomy that had so long tolerated a defective Uranian theory when, as Hanson points out, everything needed for the discovery of the unknown planet was "well-known 15 years before Le Verrier's coup."[37] The discrepancy between the actual and tabulated places of Uranus by 1845 had, after all, grown to the "intolerable quantity" of nearly two arc minutes—a little more than half the distance between the two principal components of the double-double star, Epsilon Lyrae, whose duplicity a very sharp eye is needed to detect without the aid of a telescope. Such an error, small as it seems by everyday standards, was unacceptably large given the refined standards reached by positional astronomy of the day, a gross error suggesting a large perturbing force, not a delicate adjustment of the planetary mechanism. The discovery of Neptune attested no more to the validity of the inverse-square law than the motion of any of that "oppression" of comets on which Challis worked through those fateful summer and autumn months of 1846, or by the least nod and wiggle in the motion of the Moon. To the contrary, "the astronomers were right," declares Pannekoek, "who pointed out that any of the hundreds of computed perturbations used in the planetary tables, . . . was as strong a demonstration, silently repeated every day, of the truth of science."[38]

Moreover, ironically, the science that had led to the great discovery was somewhat less exact than it had seemed to the lay mind in that fateful year, 1846. "The singularly close agreement between the observed and computed places of the planet was accidental," argues Loomis. "So exact a coincidence could not have been reasonably anticipated. If the planet had been found even ten degrees from what Le Verrier assigned as its most probable place, this discrepancy would have surprised no astronomer. The discovery would still have been one of the most remarkable events in the history of astronomy."[39]

As one dispute burned out, another flared up. This was more technical, but no less heated and of much longer duration. It concerned the very validity of Adams's and Le Verrier's calculations. It was naturally assumed about the planet, having been discovered by means of certain predicted elements, so close to its predicted place, that those elements must be extremely near the truth, and that the planet would perfectly explain those effects by means of which its own existence was detected. In fact, the planet was not moving as expected according to the orbits predicted by Adams and Le Verrier.

That this was the case became obvious as soon as the orbit of Neptune was worked out. Challis's records of August 1846 and subsequent observations up to October 13 allowed a first calculation by Adams himself that indicated the distance from the Sun was much less than expected, at just 30 AU. Challis's observations at Cambridge between October 1846 and January 1847 led to a somewhat better set of elements.[40] Then the planet disappeared into the glare of the Sun. So far it had been observed through a very short arc of its orbit, prompting the search for older "accidental" records, such as those that had surfaced in the aftermath of William Herschel's discovery of Uranus. Cacciatore's and Wartmann's objects were reexamined, in the hopes that one or the other or both might prove to be the planet. Neither was; in fact Wartmann had simply "discovered" Uranus. But an American astronomer, Sears Cook Walker, of the U.S. Coast Survey, located a possible prediscovery position in Joseph Jérôme Lalande's *Histoire Céleste Français*. The original manuscripts at the Paris Observatory were consulted; it turned out that Neptune had indeed been recorded—not once, but twice—by Lalande's nephew Michel de Lalande on May 8 and 10, 1795. Other prediscovery observations were subsequently recovered; the Scottish astronomer John Lamont, who spent most of his career at the Munich Observatory, had made uncognizant observations of Neptune on October 25, 1845, and again on September 7 and 11, 1846. Even Sir John Herschel had passed over it once, on July 14, 1830. Most remarkably, it had been recorded by Galileo himself at the very dawn of the telescopic era, in December 1612 and January 1613. When he first came across it, it was near a stationary point (changing the direction of its motion from retrograde to direct) and very close to Jupiter, whose satellites Galileo was busily observing at the time. Galileo even noted Neptune's position on a sketch on January 27, 1613, relative to the satellites and another star, and a night later entered the suspicion that the "two stars" appeared slightly further apart, but failed to check his suspicions.[41]

Walker used Lalande's records to calculate a new orbit for Neptune, which is compared with those of the hypothetical planets of Adams and Le Verrier in the table at the top of page 121.

Clearly, Neptune did not fit at all well into the Bode progression. The planet lay well inside Bode's law distance, which henceforth ceased to be regarded as a practical means of assisting future planet-seekers. Moreover, the dissimilarity of Neptune's orbit to

Elements	Walker	Le Verrier	Adams
Mean distance, a	30.25	36.15	37.25
Eccentricity, e	0.00884	0.10761	0.12062
Inclination, i	1° 54' 54"	–	–
Longitude of ascending node, Ω	131° 17' 38"	–	–
Longitude of perihelion, $\tilde{\omega}$	0° 12' 25"	284° 45'	299° 11'
Period in years, T	166.381	217.387	227.3
Mass, Sun=1	0.00006	0.000173	0.00010
Longitude, ϵ Jan. 1, 1847	328° 7' 57"	326° 32'	329° 57'

those of Adams and Le Verrier led Benjamin Peirce of Harvard to announce a further, indeed a most startling, conclusion: The Neptune discovered by Galle was not the planet computed, and its discovery near its reputed place had been a "happy accident."[42] He argued that Neptune's position was just inside a point of commensurability—its period was almost exactly twice that of Uranus. This point of commensurability created a great divide, a zone of instability, since the 2:1 ratios of the periods would lead to very considerable reciprocal disturbances between the two planets, similar to those producing the "Great Inequality" of Jupiter and Saturn that had been so triumphantly explained by Laplace. Thus, the actual behavior of Neptune, over long periods of time, would be of a character completely different from that to be expected for Adams's and Le Verrier's hypothetical planets.

Peirce's arguments were taken up a year later by Jacques Babinet, a member of the Académie des Sciences, who went so far as to suggest Adams and Le Verrier had discovered nothing at all, and that there were two perturbing planets—Neptune itself, and another planet farther beyond, which he named Hypèrion. The latter's mass, he suggested rather naively, was simply equal to the difference between that of Neptune and the planet that had been predicted by Le Verrier.[43] Not surprisingly Le Verrier himself made vigorous attempts to refute Peirce's "happy accident" theory (Adams, characteristically, remained silent), but his arguments amounted largely to quibbles and did not penetrate the core of Peirce's analysis.

Sir John Herschel discovered a way out of the impasse. Peirce had started with a "misconception of the nature of the problem."[44] The exact (as opposed to approximate) elements of the planet were of secondary importance to the prediction of the planet's position. What was of vital importance was the fact that Uranus and Neptune had been in conjunction late in 1821. This had produced a maximum perturbation that allowed Adams and Le Verrier to predict Neptune's longitude (and had also, incidentally, assured the failure of Bouvard's ill-starred *Tables*, which had appeared at the worst possible time—in 1821!). At the time of Flamsteed's observation in 1690, Herschel noted that Neptune

> must have been effectually out of reach of any perturbative influence worth considering, and so it remained during the whole interval from thence to 1800. From that time the effect of perturbation began to become sensible, about 1805 prominent, and in 1820 had nearly reached its maximum. At this epoch an alarm was sounded. . . . But the time for discussing its cause with any prospect of success was not yet come. Every thing turns upon the precise determination of the epoch of the maximum. . . . Until the lapse of some years from 1822 it would have been impossible to have fixed that epoch with any certainty, and as respects the law of degradation and total arc of longitude over which the sensible perturbations extend, we are hardly yet arrived at a period when this can be said to be completely determinable from observation alone. In all this we see nothing of accident, unless it be accidental that . . . we live in an age when astronomy has reached that perfection, and its cultivators exercise that vigilance which neither permit such an event, nor its scientific importance, to pass unnoticed. The blossom had been watched with interest in its development, and the fruit was gathered in the very moment of maturity.[45]

Much later, in 1876, Adams responded to Peirce's criticisms along somewhat similar lines. "It is true," he wrote, "that if we wished to represent the perturbations of Uranus caused by an exterior planet, through two or several synodic periods, it would be necessary to adopt a period approximately equal to the planet's perturbations." Then Peirce's objections would indeed be completely valid. However, this was not the problem he and Le Verrier had set themselves. Theirs was to represent the perturbations during only a fraction of a synodic period, in which case the approximate commensurability in the periods of Uranus and Neptune becomes irrelevant. Quoting Adams again:

> The perturbations for this limited time interval could be represented approximately by any orbit, provided only that the magnitude and direction of the perturbations exerted by the real and hypothetical planets coincided, to a sufficient degree, during the period when the perturbations reached a maximum—that is, when the planets were near their conjunction.[46]

We now know Adams and Le Verrier overestimated the assumed distances of their hypothetical planets, but their error was in turn compensated by the large eccentricity of their orbits (0.1 instead of 0.01) and excessive values for the mass. Thus between about 1790 and 1850—that is, during the very time in which the perturbations of Neptune on Uranus were perceptible—the three planets, "Adams," "Le Verrier," and Neptune itself, occupied nearly the same positions relative to Uranus, and to all intents and purposes led to the same perturbations.

Incidentally, the planets "Adams" and "Le Verrier," had they actually existed, would soon have lagged far behind the position of Neptune. At the time of writing (December 1995), "Adams" would have been near the star γ Virginis, "Le Verrier" near α Virginis (Spica), while the actual planet Neptune lies far off from either—in eastern Sagittarius, and still within 5° of Uranus, which it passed in 1993 in its first conjunction with that planet since 1821–1822. It is sobering to realize Neptune has yet to complete one circuit of the Sun since its discovery. Its period of revolution—the Neptunian "year"—is 165 Earth years. The first "Neptunian anniversary" of its discovery will not occur until 2011, when the planet will finally return to the same star field where Galle found it in 1846.

After the discovery of Neptune, Adams, who was just 27 at the time, refused a knighthood and settled into a quiet, donnish existence at Cambridge, which was congenial to someone of his shy manner, and eventually supplanted Challis as director of the Cambridge Observatory. The motion of the Moon was his speciality.

Le Verrier, the "first astronomer of the Age,"[47] was named preceptor to the Compte de Paris, received a chair in celestial mechanics, created for his benefit, at the Faculty of Sciences, and was appointed an adjunct astronomer to the Bureau of Longitudes. He was 35 at the time of the discovery of Neptune, still a young man. His intellect was restless, and his triumph, for all the honors that had been showered upon him, incomplete. Despite Neptune having been captured in the chain of his analysis, he had some unfinished business: Mercury, after all, was still off course.

THE ANOMALOUS ADVANCE OF MERCURY'S PERIHELION

LE VERRIER'S CAMPAIGN

nine

Le Verrier's Unfinished Business— Mercury

In 1848, the winds of revolution were blowing once again. King Louis Philippe, who had reigned since the July revolution of 1830, failed to take notice. His "bourgeois" monarchy had become increasingly authoritarian and out of touch, his sole advice to the poor being to get rich. On February 23, a revolt broke out in several places in Paris. By morning, the rebels were in control of the area from the quais to the boulevards, on both sides of the rue Saint-Denis. The director of the Paris Observatory, old François Arago, was an ardent republican who had known Louis Philippe in the days when he had posed as a Jacobin. He now fought by the side of the working men against the soldiers of the king. The outcome was decided when Louis Philippe called the National Guard to arms. Instead of turning on the rebels, they immediately placed themselves in the army's way, crying "Vive la Réforme!" preventing them from charging. After learning what had taken place in Paris, Louis Philippe, disguised as a Norman bourgeois, made his way in a farmer's cart to Honfleur and boarded a small boat for England.

On the evening of the victory, a howling mob invaded the Chamber of the Deputies. A provisional government was hastily formed at the Hôtel de Ville, with Arago taking a leading role. But the Second Republic proved to be short lived. Against a background of social unrest, including riots and a full-scale insurrection of the Parisian workers in June, a solid conservative majority swept into power in the elections of May 1849. Louis-Napoleon, nephew of the late emperor, was elected president of the Assembly.

Le Verrier, politically ambitious, was part of that conservative majority. His politics were diametrically opposed to those of Arago. He stood squarely on the side of order and stability. In December 1851, he supported Louis-Napoleon's coup d'état, which brought an end to the Second Republic and reestablished the Empire. For his support, Le Verrier was rewarded with a seat in the aristocratic Senate, while Arago resigned rather than take the oath of allegiance to Louis-Napoleon, henceforth known as Napoleon III. However, Napoleon made it clear that the old man should not be disturbed, but should be allowed to say and do as he liked. Arago died in 1853. In January of the following year, Le Verrier succeeded him as director of the Paris Observatory.

This was a rather strange appointment for one disinterested in the physical condition of the heavenly bodies. Flammarion tells us, "I asked him if he thought that the other planets might be inhabited like ours, what might be especially the strange vital conditions of a world separated from the sun by the distance of Neptune."[1] But his replies showed always he was uninterested in such questions. What use did he have for the conjectures of the telescope? Indeed it was Flammarion's opinion that Le Verrier never even bothered to look at Neptune through a telescope. This despite the fact that, as we now know, it was well within reach of rather modest optical means. For that matter, Adams was equally guilty.

The Paris Observatory over which Le Verrier presided was an illustrious institution. Founded in 1667 by Louis XIV, it was intended by its architect Claude Perrault (who also designed the colonnade of the Louvre) to glorify the king as much as to serve the interests of astronomy. Its facade was that of a magnificent palace. Unfortunately, the astronomers sometimes had reason to complain; the largest sums were not always spent on what was most useful—instruments for scientific research.

Indeed, unlike the Royal Observatory at Greenwich, chartered by Charles II in 1675 for the purpose of "finding out the longitude of places for effecting navigation and astronomy," the Paris Observatory lacked from its inception a well-defined object of studies. The English Astronomers Royal, beginning with Flamsteed, despite the limited and sometimes primitive resources at their disposal, had never neglected the purpose of their charter to refine the celestial movements and perfect the positions of the stars. The astronomers of Paris, on the other hand, "had no general program imposed upon them; rather they vied with each other, according to his convenience or ability or the inspiration of the moment, for the favor of the king."[2]

The man who succeeded better than any other in winning the favor of Louis XIV was the Italian-born astronomer Giovanni Domenico Cassini. He had been born at Perinaldo, near Nice, in 1629, and during the 1660s worked at the private observatory of the Marquis Malvasia at Bologna where he became famous for his observations of the rotations of the planets. His tables of the satellites of Jupiter (which he referred to, incidentally, by the names Pallas, Juno, Themis, and Ceres, rather than those by which they are known today) were published in 1668. Although based on empirical equations that he did not divulge, they were accurate and persuaded Louis to invite him to France. He arrived there in 1669, and 2 years later took up residence in the new observatory. He was a thoroughly Baroque figure—less profound perhaps than his colleague Jean Picard, but possessed of the ability to make discoveries that appealed to Louis XIV's imagination. He discovered a new satellite of Saturn in 1671, then two more in 1684, and proposed to name them the "Louisian Stars." "He was enthusiastic, good-natured, and had the power to excite Louis's curiosity and then to satisfy it; these were his objects, and he achieved them with marvellous assurance."[3]

Cassini opposed Olaus Roemer's important discovery of the finite speed of light, based on the latter's observations of the satellites of Jupiter made at the Paris Observatory in the 1670s. He also failed to grasp the innovations of the Newtonian system. Instead he identified himself with views favorable to the politics of the time, remaining an opponent of the Copernican system and a loyal adherent of the Roman Catholic faith; when he died, in 1712, the last rites were performed by the papal astronomer himself, Francesco Bianchini.

After Cassini's death, the Paris observatory remained under the control of his son, Jacques, from 1712 until the latter's death in 1756; then of his grandson César François Cassini de Thury, from 1756 until his death in 1784; and finally of his great-grandson, Jacques Dominique de Cassini (also known as Cassini IV). The Cassini dynasty, which had long served the interests of the ancien régime, was finally dissolved during year II of the Republic (1793); Cassini IV was ousted, and Jean Perny and Alexis Bouvard were placed in charge.

During the early 19th century, the observatory was a rather inefficiently run organization. By then, much of the equipment was woefully out of date—the mural-circle, for instance, was so poorly constructed that Arago was unable to publish any of his observations made with it. An exception was the fine 9½-inch Lerebours refractor, installed in 1823.

Arago began, but did not complete the process of reformation. His successor initiated his régime by personally acquainting himself with the routine work of assistants, and within a few months had laid down a system of "inflexible routine" similar to that employed by Airy at Greenwich—a not entirely happy model, since Airy's approach was later to be criticized for its gradgrindian severity.[4] It appealed to Le Verrier, who was authoritarian at heart and saw advantages in bringing every aspect of the workings of the observatory under his personal control. His assistants' routine was defined on a form prepared weekly after the model of that in use at Greenwich. "He was deeply cognizant of his indebtedness to that establishment for the greater portion of the reduced observations which served him so well in his planetary investigations; and his object was . . . to follow a system which had produced such an extended series of observations in a state ready for the mathematician."[5]

A glimpse, admittedly not unbiased, of what it was like to work at the Imperial Observatory under Le Verrier's administration is furnished in the writings of a onetime assistant, Camille Flammarion. Flammarion was born in 1842 at Montigny-le-Roi in the département of Haute-Marne. A child prodigy, by the time he was 16 he had produced an impressive manuscript that he later published in revised form under the title *Le Monde avant la création de l'homme* (The World before the creation of man). A physician called to Flammarion's home to treat him for an illness was intrigued by it, read it, and brought its young author to the attention of Le Verrier, who duly hired him as an assistant.[6]

That was in 1858. Flammarion recalled his feelings when he first arrived at the observatory:

I entered it as though it were a temple, or a holy sanctuary, with all the innocent ardor of the neophyte. There was not the least shade of skepticism; and despite my poverty, I did not think in terms of present payment or future gain. I lived alone for science; wanted to grasp, study, research, discover and contemplate the splendors of the sky, fly across the plains of the heavens, visit the other worlds it contains with the marvelous instruments of human genius. . . . To me, this seemed to be absolute happiness.[7]

His expectations of Le Verrier were equally romantic:

Since my childhood I had seen the name of the illustrious astronomer recorded on maps of the sky, for this planet had borne his name for several years, in the classic books, before it was called Neptune. This scientist, this genius, who had discovered a star with the tip of his pen, without other instrument than the strength of his calculations alone, and had boldly rolled back the frontiers of the System of the World, this man was for me a kind of saint, an inhabitant of the heavens.[8]

But he was soon disillusioned with the grand homme. At their first meeting, Le Verrier assigned Flammarion an elementary mathematics exam. When he passed it, Flammarion was brought to Le Verrier's large office and told, simply, "Sir, you start here Monday. Farewell and to your work." The work of an assistant in those days consisted of standing at a desk 7 hours a day, from 8 until noon and then from 1 o'clock until 4, doing laborious computations. Flammarion was assigned to correcting the apparent positions of the stars observed with the meridian-circle for the effects of atmospheric refraction. To do so, he took columns of numbers and added or subtracted them from the observations. Flammarion recalled:

> After the first week, one manages to carry out this work almost unconsciously—and while thinking of something else. . . . From my first days of service, I perceived that of my five colleagues, none cared the least for astronomy; none was interested in celestial contemplations, none asked what was in other worlds, none travelled in mind through the infinite spaces of the sky. . . . Excellent civil servants, accurate calculators, they saw nothing beyond the maze of numbers; it was as if they belonged to a branch of military service—they were involved only in the punctual execution of administrative orders. But all this was as Le Verrier wished it.[9]

Le Verrier, indeed, was a man of single-minded purpose (but then perhaps all true greatness is in part monomania). A fanatic for order, both in his politics and in his administration, he was intent on extending planetary theory to the limits of gravitational mechanics. The discovery of Neptune had been his grand achievement; he was not about to change his methods. To the contrary, he dreamed of reaching ever further in the direction that had led to his triumph, of pushing Newtonian gravitational theory to its limits until the last flaw had been removed from the system.

He seemed a grim man to his subordinates, with cold steely eyes and rigid inflexible features. His familiarity with them was not based on his recognition of a need to encourage their efforts and lift their flagging spirits. Their work was narrowly and rigidly defined. The observers were employed only in noting the precise instant in which stars transited the meridian, in measuring the positions of the fundamental stars that served as the framework for positional astronomy. Le Verrier cared only for the positions of the Sun, Moon, and planets relative to this framework, nothing at all for the burning questions that had fired Arago—and would fire Flammarion—of the physical constitution of these bodies. The calculator's job in turn was simply to reduce these observations, with Le Verrier himself providing the direction and supervision of the most important reductions.

"The mission of astronomy," Flammarion would protest, "ought not to end with the measurement of the positions of stars, but continue on to research into their nature. The science of the universe cannot possibly consist merely of columns of logarithms; the worlds therein are not mere inert points suspended in space. They are sources of light, warmth and life deserving to be studied." Instead of what he perceived as the dry aridity of Le Verrier's régime, a tedious enterprise of calculation and routine, Flammarion took his inspiration from Arago's posthumously published *Astronomie Populaire*, where he found "magnificent flights into the infinite, descriptions of the trains of comets, their metamorphoses, observations of the solar spots, landscapes of the Moon, the geography of Mars, the bands of Jupiter, the rings of Saturn, double stars."[10] These were the subjects of which Flammarion dreamed; Le Verrier, Arago's protégé, was insensible to such aims.

Le Verrier was a man who, as Samuel Johnson said of Milton, "was born for arduous undertakings."[11] He set himself and his assistants an "almost appalling programme."[12] Already by 1849, he had carried out another brief reconnaissance of the Mercury problem. Whereas his intelligence of 1843 had given no surety of success, he now had supreme confidence in the validity of the inverse-square law, and hoped Mercury would admit of a solution on the same lines as Uranus: "If the tables [of Mercury's positions] do not strictly agree with the group of observations, we will never again be tempted into charging the law of universal gravitation with inadequacy," and borrowing a page from the Uranus saga, he intimated the solution might involve "some material cause whose existence has escaped us,"[13] implying, in other words, a missing mass.

For the moment he carefully avoided dogmatism. A good deal of ground had to be worked over before he could reach any definite conclusions. Above all, the theory of the motion of the Sun was in need of revision; this had been evident to him since 1843. This was obviously of critical importance, since Mercury's measured positions at its transits were all made relative to that body. Systematic errors could not yet be ruled out, so he began from first principles. First he attempted a revision of the catalogs of the fundamental stars, so as to rule out any mistakes from that quarter.

This finished, he (or rather his overworked computers) undertook the backbreaking labor of reducing some 9,000 meridian-circle observations of the Sun taken at Greenwich between 1750 and his own time. However, these were found to be less reliable than he expected; some discrepancies attributed to theory were, he concluded, due to observational errors.

He published his massive reductions of these observations in 1852. Then other investigations intervened, including, in 1855, a thorough reconsideration of the problem of the stability of the solar system, which he published the following year as the second volume of the newly founded *Annales de l'Observatoire Impérial de Paris*, under his exclusive editorship. He carried this important investigation to much higher terms than Laplace had done, and concluded his great predecessor's approximation was insufficient to establish the long-term stability of the solar system. However, he remained uncertain whether the system of smaller planets, composed of Mercury, Venus, Earth, and Mars, would enjoy stability indefinitely. Astronomy, as Jacques Laskar has noted, had reached a stalemate.[14]

After a brief digression into the theory of the comets, Le Verrier published his revised tables of the Sun in 1858, forming the fourth volume of his *Annales*. His most important result was a correction of 5″ to the mean motion of the Sun, which amounted to a decrease of the Earth–Sun distance from 95,000,000 miles (the value Encke had found from his reduction of the 1761 and 1769 Venus transit observations) to 92,500,000 miles. He also increased the mass of the Earth by 10% and that of Mars by 10%. By making these changes, Le Verrier found he was able to reconcile discrepancies in the theories of Venus, Earth, and Mars. In a way, observes Zachary, his whole method suggests a strange metaphysical disposition toward the primacy of mathematics, that all knowledge of the universe is encapsulated by the stroke of a pen.[15] No greater illustration of this existed than the certainty with which Le Verrier predicted Neptune.

At last Le Verrier was ready to turn to the next step in the unfolding of his grand design—the working out of each planet's mo-

tion one at a time, beginning with Mercury. He had already announced his basic plan of attack in the volume on the Sun:

> (i) After having determined the perturbations and assembled the
> formulas which represent the coordinates of the planet, I will (ii)
> discuss the observations which will be used. (iii) The comparison of
> these observations with the theory will lead to the rectification of
> the values adopted for the elements which enter into the formulas.
> I will finish up with (iv) the tables necessary to calculate the posi-
> tion of the planet.[16]

The problem of Mercury as defined by Le Verrier in 1859 was
different from that he faced in 1843. He had eliminated the potential sources of error outlined in his earlier attempt only to disclose
unacceptable discrepancies.

The observational material he attempted to reduce consisted
of a mass of meridian-circle observations and the best-observed
contact points recorded at 14 transits of Mercury—nine November
transits between 1697 and 1848, and five May transits between
1753 and 1845. In practice, he relied almost entirely on the transit
observations, since although few in number, they were unsur-
passably precise—"provided only that the place where the observation was made is well known," Le Verrier pointed out, "and that
the astronomer was equipped with a good enough telescope, and
that his clock was accurately regulated to within a few seconds,
the timing of the internal contacts allow the position of the center
of the planet relative to the Sun to be measured to within an error
of one second of arc."[17]

At the November transits (those taking place at the planet's
ascending node), the residuals between theory and observation
were within this margin of error. However, not so for the May
transits (those occurring at the descending node); here Le Verrier
encountered a problem. Working back from the 1845 transit, he
noted a progressive error from one transit to the next, which
reached 12 arc seconds for that of 1753. The errors could not be due
to any shortcomings of the theory of the Sun; he had now settled
that. Nor could they be attributed to errors of observation, since
this would be "to suppose that all astronomers have made great

mistakes in measuring the times of the contacts. These mistakes, moreover, would have had to vary progressively in time, and differ by several minutes at the end of the period of 92 years. *Chose impossible!*"[18] The only resort, and Le Verrier well knew it, was to modify the values in the time-dependent elements of the orbit.

In order to determine the necessary modification, Le Verrier set up equations of condition in terms of a correction of the longitude of the ascending node. What was implied, he found, was an additional increment to the rate of advance of the planet's perihelion. For the same reason as the case of the Moon's apsides, the perihelion of Mercury slowly advances its position, so that over time its elliptic orbit traces out a rosette pattern in space (Figure 14). In producing these perturbations, Venus, Mercury's nearest neighbor, is the main culprit, but the other planets also contribute in greater or lesser degree. The expected increments due to each planet Le Verrier calculated as follows, with the results given in arc seconds per century:[19]

Venus	280.6"
Earth	83.6"
Mars	2.6"
Jupiter	152.6"
Saturn	7.2"
Uranus	0.1"
Total per century	526.7"

Note that massive Jupiter contributes over a quarter to the total rate of advance, despite its great distance from the Sun.

The planetary perturbations led to an expected rate of advance of the perihelion of Mercury of 1¼" per orbit, or 527" per century. In order to account for the transit observations, Le Verrier declared it was necessary to suppose that "in a century Mercury's perihelion turns not merely 527" as a result of the combined actions of the other planets, as [Newtonian theory] requires, but rather 527" + 38". There is, then, with the perihelion of Mercury, a progressive displacement reaching 38" per century, and this is not explained."[20] Here Le Verrier announced a discovery as conse-

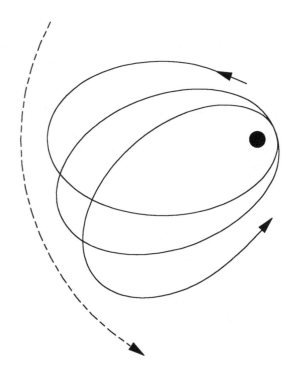

14. The anomalous advance of Mercury's orbit, greatly exaggerated. The ellipse advances 43 arc seconds per century.

quential—and in the long run arguably more so—than that of Neptune; he had identified the famous anomalous advance in the perihelion of Mercury as an outstanding problem of science.

Assume this additional displacement of the perihelion of Mercury, says Le Verrier, and the observations of Lacaille, Delisle, Bougour, and the Cassinis all fall into place to within 1 arc second— in some cases to within 0.5 arc seconds. Fail to assume it, and there is no choice but to accuse all these skillful observers of the most egregious errors. As with Uranus, Le Verrier refused to impugn the accuracy of these skilled observers; he had no doubt the effect revealed by the observations was real. Thus he concluded: "Such an advance of the perihelion of Mercury is indispensable."[21] But recognizing the anomalous advance was one thing, explaining it another.

Le Verrier approached the problem as a conservative, trying once more to rescue Newtonian theory from its own resources. Seen from this perspective, the anomalous advance of the perihelion of Mercury must, he argued, "be the result of masses still unknown." An increase of 10% in the mass of Venus would do; he realized this created more problems than it solved, since Venus's perturbations of the Earth's motions then became unacceptably large. His pursuit of this unsuccessful stratagem at least narrowed the problem. The missing mass, whatever form it took, must be such that it did not sensibly affect the motion of the Earth:

> We do not know whether it would affect Venus; until this point has been clarified, we shall say that this action may be imperceptible or at least weaker than on Mercury. On this hypothesis, the sought-for mass should be found within Mercury's orbit. If, in addition, its orbit is not to intersect with Mercury's, its distance at aphelion must never exceed 8/10 of Mercury's mean distance—i.e., 3/10 the mean distance from Earth to Sun. Our observations of Mercury . . . have shown no variation in the inclination of its orbit . . . [so] the orbit of the disturbing mass is [not] much inclined to Mercury's. . . . The orbit . . . has only a tiny eccentricity . . . the perturbing mass must be reckoned the more considerable the nearer the Sun we place it.[22]

Now Le Verrier played his trump card—irresistible to the discoverer of the unknown mass, Neptune, viz. there must be an additional mass between Mercury and the Sun. The requirements were that it had to account for the advance of Mercury's perihelion, yet at the same time produce no marked effect on the planet's nodes or on the motions of either Venus or Earth. Assume, for the sake of argument, there was an inner planet. The rapid motion would average out (for long-term effects) to something quite like an equatorial bulge on the Sun. It would effectively add a little mass to the Sun, which would be absorbed in all determinations of the Sun's mass from astronomical observations, and would also add a $1/r^3$ force, able to produce the observed advance of the perihelion as well as a very slow advance of the line of the apsides (the latter, however, is much harder to measure accurately for a planet in an orbit so near the plane of the ecliptic as Mercury).[23]

The missing mass hypothesis was attractive. Was it a planet? Not necessarily; there were compelling reasons to think otherwise. "At the mean distance of 0.17 [AU]," Le Verrier wrote, "the perturbing mass would be equal to that of Mercury. Its greatest deviation [from the Sun] would be just under 10°. This planet would shine more brilliantly than Mercury [since the brightness of a body increases by its proximity to the Sun]. . . . How could a planet, extremely bright and always near the Sun, fail to have been recognized during a total eclipse? And would not such a planet pass between the Sun and Earth, thereby making its presence known?"[24]

While a single planet was admittedly unlikely, Le Verrier glimpsed an alternative. "All these difficulties disappear," he wrote, "if instead of a single planet, we invoke the existence of a series of small bodies [*corpuscles*] orbiting between Mercury and the Sun."[25] In Le Verrier's view, it was essential to admit their existence. Such bodies might be unobservable.

Le Verrier published his hypothesis September 12, 1859, in a letter to his colleague Hervé Faye, the secretary of the Paris Académie des Sciences. If the small bodies he hypothesized actually existed, Le Verrier told Faye, some might be larger than others. He included this plea to astronomers:

> We cannot establish their existence other than by observing their motion across the Sun's disk; this discussion should make astronomers the more zealous in studying each day the Sun's surface. It is important that every regular spot . . . however tiny, be carefully followed for a few months to determine its nature through familiarity with its motion.[26]

Faye appended his own endorsement of Le Verrier's ideas, hinting the reason intramercurial bodies had not been found was because no one had bothered to look for them. In his view, the best time to search would be during a total eclipse of the Sun. The next one would be visible over Spain July 16, 1860. Moreover, a search conducted with the aid of photography, just then being successfully applied for the first time for astronomical purposes, would be preferable to a hurried (and unreliable) visual search.

Faye's claim that intramercurial bodies had not previously been reported was wide of the mark. The Sun's face had been subject to long and careful examination since the first application of the telescope in the 1610s by astronomers such as Galileo, Fabricius, and Scheiner. Although much of the late 17th-and early 18th-centuries records are blank—partly because of the so-called Maunder minimum, a relative paucity of sunspot activity that occurred between about 1645 and 1715—by the 1780s, the energetic J. H. Schroeter had stimulated renewed interest. In 1789, he published a book, *Beobachtungen über die Sonnenfackeln und Sonnenflecken* (Observations of the Solar Faculae and Sunspots).[27] As with so much of Schroeter's work, it was less important in its own right than in its influence on others. The Lilienthal astronomer was firmly convinced sunspots had independent motion, and that a few moved with unusual rapidity. He himself provided a few observations of this sort, but others, inspired by him, produced more numerous records. Thus Pastor Fritsch, at Magdeburg, observed rapidly moving spots March 20, 1800; February 7, 1802; and October 10 , 1802. On various occasions in the 1820s and 1830s, so did another German astronomer, J. W. Pastorff.

The question had not been settled when Heinrich Schwabe, a Dessau pharmacist, entered the field in 1826. His interest in the Sun was stimulated by Karl Harding, Schroeter's onetime assistant at Lilienthal. "It was in consequence of a suggestion of my old and valued friend Harding of Göttingen," Schwabe recalled, "who wrote to me that there was a great want of physical observations of the Sun, that the subject presented an almost unworked field, and that this labor might be rewarded by the discovery of a planet interior to Mercury."[28] Whenever the clouds permitted, Schwabe maintained a regular daily watch of the solar disk from Dessau, an average of 300 days a year. He first concluded Schroeter's hypothesis was unfounded; instead, he found, "the apparent movement was attributable rather to the enlarging of one side and the filling in of the other side of a nuclear spot, whereby the estimated center was shifted."[29] Continuing to maintain careful records of what he saw on the face of the Sun, Schwabe, after 12 years, announced he had failed to find any planets, a singularly important negative result. He had, however, found something even more important—the sunspot cycle. Thus he wrote: "I may compare myself to Saul, who went out to seek his father's asses, and found a kingdom."[30] Schwabe was an admirably cautious observer; he did not immediately publish his findings but spent 6 more years satisfying himself, then another 13 convincing other astronomers of its truth.

Had any intramercurial planet existed, it would have been hard for Schwabe, during a vigil of 30 years, to have missed it. Nor was Schwabe alone. Another searcher was J. F. Benzenberg, who dedicated a large refractor to the purpose at his private observatory at Bilk, near Düsseldorf. Benzenberg began searching in 1834, and was still doing so in 1845 when he hired as an assistant a promising young observer, J. F. Julius Schmidt. Since Benzenberg reserved the large refractor entirely for his personal use in intramercurial planet search, keeping it under lock and key at all other times, Schmidt did not remain with him for long.[31]

Meanwhile, in America, Edward Herrick, Yale librarian, made curious by some of Pastorff's records from the 1830s, had started his own search. He had scarcely got underway when Le Verrier published his 1845 tables of Mercury, whose accuracy made him doubt the existence of such planets. After Neptune was discovered, Herrick's interest revived. With his friend, Francis Bradley, he began a regular scrutiny of the Sun's disk in 1847. Herrick and Bradley also tried to explore the Sun's surroundings with a telescope that had a blackened interior tube. They found nothing. Herrick wrote to Le Verrier in 1859: "It is remarkable that during the last twenty years we have no observation by Schwabe or others of the transit of such bodies across the Sun." Nevertheless, Herrick cautioned, "it may be that the orbit of the planet has a high inclination."[32]

Also in 1859, the Zurich astronomer Rudolf Wolf wrote more encouragingly to Le Verrier. A keen student of sunspots, Wolf had scouted out many of the earlier records of sunspots with unusually rapid motions. In his view they seemed a likely indication of intramercurial bodies. He drew up a list of them, which he placed at Le Verrier's disposal.[33]

Every concept has its prophets and forerunners; Le Verrier's hidden planet hypothesis is no less than any other. "It is not improbable that a planet may exist . . . between the orbit of Mercury and the Sun," surmised Thomas Dick, the popular Scots science writer, in 1838.[34] Fourteen years later American mathematician Benjamin Peirce intimated "there are some indications of the secular action of a planet within the orbit of Mercury."[35] That same year another American, Daniel Kirkwood, added his contribution with his "Analogy," an empirical law that by its implied connection with Laplace's nebular hypothesis promoted the existence of unobserved debris in the solar system, a possibility that encouraged the hope of material inside the orbit of Mercury.[36]

Jacques Babinet accounted for the apparent discordance between the predicted and observed elements of Neptune by assuming the existence of yet another exterior planet, Hypèrion. He had earlier speculated that prominences near the Sun seen during the 1842 eclipse were due to "incandescent clouds of a planetary kind circling the Sun in the form of a train or portions of a ring." He had gone so far as to suggest the mass should be named *Vulcan*, while other analogous masses might be called the Cyclopes.[37] Not surprisingly, his ideas met with skepticism. In Babinet's defense, it must be recalled solar prominences were a complete mystery at the time—Faye, for instance, regarded them as nothing more than illusions! Nor can we overlook the ring of intramercurial meteoroids cited by the British physicist William Thomson (Lord Kelvin) in his widely published 1854 paper "On the Mechanical Energies of the Solar System." Another interesting speculation was offered by a Dutch meteorologist, C. H. D. Buys-Ballot, who from supposed variations in the temperature of the Earth's atmosphere inferred the periodic interposition of a ring or ring fragments located between the Earth and the Sun. This was a hypothesis briefly recalled at the time Le Verrier published his own hypothesis.[38]

Eighteen fifty-nine, the year of Le Verrier's hidden planet hypothesis, was the year Oregon joined the Union and John Brown seized Harper's Ferry; it was the year of Garibaldi and the Risorgimento. Above all, it would be Charles Darwin's year. It was also when Kirchhoff and Bunsen established the principles of spectroscopy, thus pointing the way to deciphering the chemical nature of the stars—something the French philosopher Auguste Comte had confidently stated a few years earlier would remain forever unknowable.

It was also the year of crisis in Newtonian mechanics. The difficulties were, in one sense, very small—the 38 arc seconds per century of the anomalous advance of Mercury's perihelion was an amount smaller than could be discerned by the naked eye. Put another way, it was less than the apparent diameter of the disk of the planet Jupiter. In another sense, however, it required the spanning of enormous chasms. Celestial mechanics was nearly perfected; "everything advances," Tisserand later wrote, "except for one or two small difficulties over which our successors will no doubt triumph."[39]

The "small difficulties" Tisserand had in mind were the so-called secular acceleration of the Moon, a gradual slowing of the Moon's motion first noted by Halley in 1693, and the anomalous advance of the perihelion of Mercury. The formidable powers of two great mathematicians who had brought the theory its grandest triumph in the discovery of Neptune in 1846 were fully concentrated in 1859 on these "small difficulties." Adams, at Cambridge, grappled with the Moon; Le Verrier, at Paris, with Mercury. Adams's problem seemed to require the invocation of tidal friction, producing a slowing of the Earth's rotation. Le Verrier posed a tentative solution to his problem along the lines of the Neptune solution—this time without the bravado of that earlier occasion. A shadow of doubt, perhaps, hung over the whole enterprise. And yet the alternative was unthinkable, since within the framework of Newtonian mechanics no other solution seemed possible. (Adams, incidentally, was in complete agreement with Le Verrier's formulation of the Mercury problem and later wrote, "the transits of the planet across the Sun furnish such accurate observations, as to leave no doubt of the reality of the phenomenon in question; and the only way of accounting for it appears to be to suppose, with M. Le Verrier, the existence of several minute planets, or of a certain quantity of diffused matter circulating about the Sun within the orbit of Mercury."[40])

At this moment the most sublime theory ever framed by the human mind seemed to hang in the balance. Suddenly there came an apparent deus ex machina to Le Verrier's dilemma. Shortly before Christmas 1859, he received a startling communication. Improbable yet seductive, it launched one of the strangest adventures in astronomical history.

ten

The Doctor's Tale—Vulcan!

Le Verrier received a startling communication from an amateur astronomer and physician, Edmond Modeste Lescarbault. Writing from Orgères-en-beauce on December 22, 1859, Lescarbault informed the imperial astronomer that the planet Le Verrier has inferred from the anomalous advance of the perihelion of Mercury actually exists. Lescarbault has already observed it in transit across the Sun on the previous March 26, 1859.

Lescarbault, who was born at Chateaudun in 1814, has been remembered as "a kindly man, a bit of a dreamer who seems to have practised more astronomy than medicine."[1] This is hardly a condemnation, since at the time doctors were often consigned to little more than watching, waiting, and hoping with their patients. He and Le Verrier had never met. Indeed they were very different kinds of men: The country doctor was completely unknown; the arrogant, aristocratic Parisian stood at the summit of the astronomical world. And yet their destiny was to be strangely intertwined.

First attracted to the possibility of other planets in 1837 by the imprecision of Bode's law, Lescarbault had been seized with the idea of searching for an unknown interior planet as he watched Mercury hurtle in front of the Sun at the transit of 1845.

The unforgettable sight rekindled the old passion. If another inner planet existed, apart from Mercury and Venus, it might, he realized, reveal itself in the same fashion. Not until 1853 did he launch an irregular and unsystematic perusal of the Sun's disk for possible planets.

His main instrument was a 3¾-inch Cauche achromatic refractor, made in 1838, which had a focal length of 4 feet 10 inches and a magnifying power of $150 \times$. It was supported by a wooden pillar with three feet, the points of which rested on a frame also with three feet, and had leveling screws. It also had a finder and a small telescope magnifying six times and was equipped with various measuring instruments of Lescarbault's own devising:

> the eyepiece of the telescope and the eyepiece of the finder telescope each had at its focus two wires crossing at right angles, and the wires of the latter were so adjusted that a star seen at their intersection was seen at the intersection of the wires of the telescope. There were also in the eyepiece of the finder two wires parallel to, and on opposite sides of, each cross-wire. . . . A circular card about 6 inches in diameter, and graduated to half degrees, was placed concentric with the tube of the eyepiece of the finder, and apparently could be moved both about the tube and, with the tube, about the axis of the finder.[2]

Lescarbault stole moments, whenever possible between patients, rushing back and forth between his surgery and his telescope to observe the Sun. In 1858, he began a more rigorous vigil, aided by the fact he had a terrace on which to set up his telescope. Shortly afterwards, a sign possibly of growing prosperity, he indulged by having it mounted upon a revolving platform under a small cupola, above the surgery. He regulated his watch by the Sun's passage across the meridian, using a small transit instrument. Adjusting the rest of his apparatus and directing the Cauche refractor to the Sun, he scrutinized the solar disk.

His quest seemed chimerical, even to himself. After all, what was the chance of a small planet escaping the attention of so many diligent observers, going back all the way to the Chaldeans and the ancient Chinese? Under the circumstances, he could hardly have been surprised when the year ended without revealing anything unusual.

A new year opened, one suspects, for Lescarbault, with no more than the usual expectations of a country doctor. But the world would later regard 1859 as an auspicious year. It was also to be Lescarbault's year.

On March 26, 1859, the sky was overcast throughout much of France. However the Sun shone brightly on Orgères. At about 4 p.m., local time, Lescarbault found a moment of leisure and, as on countless other occasions, turned his Cauche achromat toward the Sun.

He noticed a small spot on the upper part of the Sun's disk: well-defined, perfectly round, perhaps a quarter the size of Mercury as he remembered seeing it with the same instrument at its transit 14 years earlier. Perhaps the speck was no more than an ordinary sunspot. But it showed movement. No sunspot would exhibit motion in so short a time. Lescarbault was of sound faculties, not prone to hallucinations, and had no reason to doubt the evidence of his senses,

> a spot like which perhaps
> Astronomer in the Sun's lucent Orb
> Through his glaz'd Optic Tube yet never saw.[3]

He quickly brought the object to the intersection of the wires of his telescope and attempted, roughly, to measure its distance from the Sun's limb and its position-angle. Since the spot was already on the Sun when he first examined it, he could only estimate the time and place of ingress or entry. But the mystery object was hardly in his ken before he was called away to attend to a patient, who threatened like the man at Porlock to interrupt his meditation (Figure 15).

After responding to the patient's complaint, he hastened back to the observatory, fearful lest the spot had vanished—a circumstance that would, of course, consign it irretrievably to the realms of unsatisfied surmise or venturesome speculation. Fortunately, the distraction was minor. The spot was still there, although its continued motion had brought it to the opposite side of its chord of transit across the Sun. Again with the aid of his rude apparatus, he measured its distance from the limb and procured another set of position angles.[4] Again he checked the time. Since his old watch

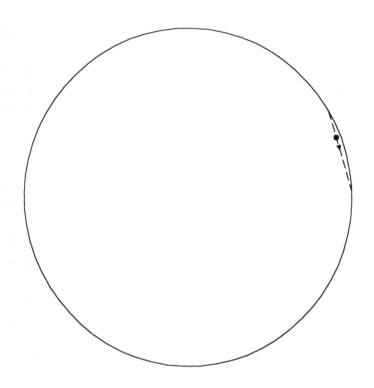

15. Lescarbault's planet on the Sun, March 26, 1859.

had only a minute hand, he beat out the seconds with a pendulum, an ivory ball suspended by a silk thread, the instrument he used to take the pulses of his patients. All told, he estimated the spot had been on the disk for 1 hour 17 minutes 9 seconds. Then, as suddenly as it came, it exited the face of the Sun.

Diffident of his abilities, naturally shy and unassuming, Lescarbault kept the observation to himself for 9 months. He wanted to see the planet again before he made a public announcement. He knew it was only by rare good luck he had seen it and that years might pass before the opportunity would be repeated. Had he been able to calculate its period, he would have had a somewhat better chance of recovering it. Still, he was no mathematician, and found the project—simple as it was—daunting. Only when he read an article in *Cosmos,* a scientific journal edited by the Abbé François Moigno, which detailed Le Verrier's work on the anomalous advance of the perihelion of Mercury, did he overcome his long hesitation in laying his approximate data before the scientific world.

Thus he came to pen his December 22 letter to Le Verrier. Its main purpose was to furnish a précis of his observations. The small black spot's entry on the solar disk occurred, he recounted, at a point 57° 22' 30" west of the upper extremity of the vertical diameter of the Sun, and at 14 seconds before 4 p.m. Orgères time (he gave an error of only 1–5 seconds in the time). The object reached its least distance from the center of the Sun at 4 hours 38 minutes 20 seconds p.m., and exited the Sun's disk at a point 85° 45' 0" to the west of the lower extremity of the Sun's vertical diameter at 5 hours 16 minutes 55 seconds p.m. The chord of its passage was very nearly tangential to the Sun's disk, but someday, he confided to Le Verrier:

> someone will again observe the transit of a perfectly round, tiny, black dot, traversing a plane-line inclined at an angle of between about 5⅓° to 7⅓°; the orbit described by this plane-line cutting that of the Earth at about 183° from south to north; and unless there is an enormous eccentricity in the black dot's orbit, it should be visible on the Sun's disk for 4½ hours. This black dot will very probably be the planet whose path I observed on March 26, 1859, and it will be possible to calculate all the elements of its orbit. I have reason to believe that its distance is less than that of Mercury, and that this body is the planet, or one of the planets, whose existence in the heavens you, Monsieur Director, had, some months ago, predicted in the neighborhood of the Sun, by the marvelous power of your calculations; which enabled you also to recognize the conditions of the existence of Neptune, and fix its place at the confines of our planetary system, and trace its path across the depths of space.[5]

Lescarbault entrusted his letter to a local colleague, M. Vallée, Honorary Inspector-General of Roads and Bridges, who carried it to the political center—Paris—and into the august presence of Le Verrier himself. The grand homme perused the letter with annoyance. He was a man burdened with heavy responsibilities, a man easily annoyed. Despite the fact the letter's author was completely unknown, Le Verrier's curiosity was aroused, and he was unable to dismiss the communication completely. "The details," he later recalled, "were such as allowed me to place in it a certain confidence."[6] On the other hand, he was not about to grant the report his unqualified belief; "though he had a secret conviction that the story might be true, yet the predominant feeling in his mind was to unmask an attempt to impose upon him, as the person more likely than any other astronomer to listen to the allegation that his prophecy had been fulfilled."[7] He was irked, too, that having made such a remarkable discovery, Lescarbault should have kept the facts to himself for 9 months—a delay that did not seem to be sufficiently accounted for by Lescarbault's comment that he had wanted to see the black spot again before making it public.

Incensed by these circumstances but anxious to learn more, Le Verrier resolved to pay Lescarbault an unannounced visit. It was Friday. Two days hence his father-in-law, Monsieur Choquet, was to hold a New Year's Day party to which many notable personalities had been invited. Le Verrier checked the railway timetable. It was just enough time to learn the truth directly from this presumptuous amateur and get back to Paris in time for the party. Accordingly that same day, December 30, accompanied by M. Vallée's son, who agreed to be a witness of the planned inquisition, he set out for Orgères. Since the nearest rail station was 12 miles from their destination, he and M. Vallée completed the last part of their journey on foot through open country.

There are two histories of what happened next: the very conservative account Le Verrier addressed to the Académie on January 2, 1860, and the rather amusing version he gave amidst the champagne and brilliance of the previous evening when the events were still fresh in his mind. It was fortunate he had among his audience on the latter occasion the celebrated scholar M. L'Abbé Moigno who reproduced the whole fascinating story as it fell from Le Verrier's lips in the next issue of his journal *Cosmos* (January 6, 1860).[8]

Arriving at the cluster of neat gray houses, Le Verrier and Vallée noticed one in particular: a single-storey, barnlike building, with a tall pitched roof, distinguished with a lofty turret at one end and two small extensions at the other. It was very plain, but sturdily built, with a tall arched door and large window of the same design. A shuttered cupola crowned the turret. Le Verrier, with his fine blonde leonine head and haughty manner, walked briskly up to the house and rapped on the door. It opened to reveal a timid, obsequious man, almost crushed by the stranger's hauteur.

"One should have seen M. Lescarbault," says Abbé Moigno, "so small, so simple, so modest, and so timid, in order to understand the emotion with which he was seized, when Le Verrier, from his great height, and with that blunt intonation which he can command, thus addressed him:

"'It is then you, sir, who pretend to have observed the intramercurial planet, and who have committed the grave offense of keeping your observation secret for nine months. I warn you I have come here with the intention of doing justice to your pretensions, and of demonstrating either that you have been dishonest or deceived. Tell me, then, unequivocally, what you have seen.'

"The lamb trembled at this rude summons from the lion, and, unable to speak, stammered out this reply: 'On the 26th March, about four o'clock, I directed my telescope to the Sun, as I had been in the habit of doing, when, to my surprise, I observed, at a small distance from its margin, a black spot well defined and perfectly round, and advancing with a very sensible motion upon the disk of the Sun. Unfortunately, however, a customer arrived. I came down from the observatory, and in this painful situation I replied as I best could to the inquiries which were made, and returned to the observatory. The round spot had continued its transit; and I saw it disappear at the opposite margin of the Sun, after having been projected upon his disk for nearly an hour and a half.'

" 'You will then have determined the time of the first and last contact,' asks Le Verrier, 'and are you aware that the observation of the first contact is one of such extreme delicacy that professional astronomers often fail in observing it?'

" 'Pardon me, sir,' replies the Doctor, 'I do not pretend to have seized the precise moment of contact. The round spot was upon the disk when I first perceived it. I measured carefully its distance from the margin, and, expecting that it would describe an equal distance, I counted the time which it took to describe this second distance, and I thus determined approximately the instant of its entry.'

" 'To count the time is easy to say, but where is your chronometer?'

" 'My chronometer is a watch with minutes, the faithful companion of my professional journeys.

'What! with that old watch, showing only minutes, dare you talk of estimating seconds? My suspicions are already too well founded.'

" 'Pardon me,' was the reply, 'I have also a pendulum which nearly beats seconds.'

" 'Show me this pendulum,' says Le Verrier. The Doctor goes upstairs into the observatory, and brings down a silk thread, to which an ivory ball was suspended. 'I am anxious to see how skillfully you can thus reckon seconds.'

"The lamb acquiesces. He fixes the upper end of the thread to a nail, and after the ivory ball has come to rest, he draws it a little from the vertical, and counts the number of oscillations corresponding with a minute on his watch, and thus proves his pendulum beats seconds.

" 'This is not enough,' replies the lion; 'it is one thing your pendulum beats seconds, but it is another that you have the sentiment of the second beaten by your pendulum in order that you may count the seconds in observing.'

" 'Shall I venture to tell you,' answered the doctor, 'that my profession is to feel pulses and count their pulsations? My pendulum puts the second in my ears, and I have no difficulty in counting several successive seconds.'

" 'This is all very well for the chapter of time,' says the Director; 'but in order to see so delicate a spot, you require a good telescope. Have you one?'

" 'Yes sir, I have succeeded, not without difficulty, privation, and suffering, to obtain for myself a telescope. After practicing much economy, I purchased from M. Cauche, an artist little known, though very clever, an object-glass nearly four inches in diameter. Knowing my enthusiasm and my poverty, he gave me the choice among several excellent ones; and as soon as I made the selection, I mounted it on a stand with all its parts; and I have recently indulged myself with a revolving platform, and a revolving roof, which will soon be in action.' Le Verrier went to the upper storey, and satisfied himself with the accuracy of the statement.

" 'This is all well,' says he, 'in so far as the observation itself is concerned; but I want to see the original memorandum which you made of it.'

" 'It is very easy,' answered the Doctor, 'to say you want it; but though this note was written on a small square of paper, which I generally throw away or burn when it is of no further use, yet it is possible I may still find it.' Running with fear to his *Connaissance des Temps,* he finds the note of the 26th March 1859 covered with grease and laudanum performing the part of a marker.

"The lion seizes it greedily, and, comparing it with the letter which M. Vallée had brought him exclaims: 'But, sir, you have falsified this observation; the time of emergence is four minutes too late.'

" 'It is,' replied the lamb. 'Have the goodness to examine it more narrowly, and you will find the four minutes is the error of my watch, regulated by sidereal time.'

" 'This is true; but how do you regulate your watch by sidereal time.'

" 'I have a small telescope—here it is—which you will find in such a state as to enable me to obtain the time to a second, or even to some fractions of a second.'

"Satisfied on this point, Le Verrier then wished to know how he determined the two angular coordinates of the points of contact, of the entry and emergence of the planet, and how he measured the chord of the arc which separates these two points. Lescarbault told him that this was reduced to the measuring of the distances of these points from the vertical, and the angles of position, which he did by systems of parallel axes, and the divided circle of cardboard placed upon his finder.

"Le Verrier next inquired if he had made any attempt to determine the object–Sun distance from the period of four hours which it required to describe an entire diameter of the Sun. The Doctor confessed to many attempts, but, not being a mathematician, had not succeeded. . . . Le Verrier asked to see the rough draft of the calculations. 'My rough drafts!' Lescarbault replied, 'Paper is rather scarce with us. I am a joiner as well as an astronomer. I calculate in my workshop, and I write upon the boards; and when I wish to use them in new calculations, I remove the old ones by planing.' On visiting the carpenter's shop, they found the board, with its lines and numbers in chalk still unobliterated."[9]

Le Verrier's cross-examination lasted in all for about an hour. (It is not clear, by the way, whether Lescarbault even guessed who his imperious visitor might be; he found out later, of course.) Le Verrier satisfied himself on every point; the doctor's 9-month silence, for instance, sprang "only from the modesty and reserve one finds in places far from the activity and agitation of cities."[10] Before leaving Orgères, Le Verrier made a few local inquiries as to Lescarbault's private character, obtaining testimonials from the village curé and juge de paix; everything seemed to be in order.

He had come to Orgères at least outwardly disdainful of the amateur's work. But his strong motivation was to yield belief to the apparent confirmation of his theory. Inwardly possessed of a secret conviction the story might now be true, he left convinced an intramercurial planet had been found and congratulated Lescarbault "on the important discovery which he had made."[11]

The first official announcement of Lescarbault's apparent discovery was made by Le Verrier at a public meeting of the Académie des Sciences on January 2, 1860, one of the strangest meetings in the history of that august body. After quoting Lescarbault's letter and the circumstances leading up to his visit, Le Verrier remarked:

We found in M. Lescarbault, a man long devoted to the study of science, surrounded with instruments, with apparatus of every description, all constructed by himself, and capped with a small rotating cupola. M. Lescarbault freely allowed us to examine, in the most scrupulous detail, every instrument used by him, patiently described its workings, and above all, went into the circumstances of the transit of the planet across the Sun.[12] . . . The explanations of M. Lescarbault, and the simplicity with which he presented them, produced on our minds an absolute conviction that the details of his observations ought to be admitted to science.[13]

So great was Le Verrier's prestige that the members of the Académie accepted the discovery with applause and acclamation. It was 1846 all over again, although this time even more gratifyingly so: A French astronomer had accomplished what another Frenchman had predicted.

The midwinter announcement took Paris by storm.[14] Le Verrier was the toast of Paris, the lion of the gayest salons. Lescarbault too received his share of honors. Le Verrier himself proposed to M. Rouland, the Minister of Public Education, that the obscure doctor receive the Legion d'Honneur; it was duly conferred upon him (in absentia) by decree of the emperor, Napoleon III, on January 25, 1860. Meanwhile, the members of the medical faculty of Paris wanted to honor him at a banquet in the Hôtel de Louvre. He was solicited by learned societies in Chartres and Blois. But Lescarbault declined all these invitations, pleading "his simple and retired habits, and the difficulty of leaving the patients under his care."[15] A timid man, he prefered to remain in the background of events.

Meanwhile, the rest of the world reacted to the news from Paris. Lescarbault's ingenuity in assembling his own instruments received favorable mention in London in *The Spectator*.[16] He was congratulated by the Royal Astronomical Society, which recognized "the singular merit of M. Lescarbault's observations . . . Astronomers of all countries will unite in applauding this second triumphant conclusion to the theoretical inquiries of M. Le Verrier."[17] In America, one of the leading Eastern dailies berated astronomers at the Harvard Observatory at Cambridge, Massachusetts, for having "permitted an amateur astronomer, armed with a telescope no more powerful than an opera glass, to make this greatest of all modern discoveries in the heavens," and added, "these gentlemen at the Cambridge Observatory, having at their command the best instruments in America, seem to have been asleep or too lazy to do anything except to draw their pay."[18]

From Lescarbault's observations, Le Verrier, consummate mathematician, carried out the calculation that had baffled the country doctor. By assuming its orbit was roughly circular, Le Verrier placed the new planet at a distance of 0.147 AU from the Sun, or some 13 million miles. It revolved around the Sun once every 19 days 17 hours, and its orbit was inclined to the ecliptic by 12° 10'. According to Le Verrier, it could never appear more than 8° from the Sun, and it must be much dimmer than Mercury for it not to have been seen before. Furthermore, transits of the planet were due to occur whenever it passed near its nodes; there should be at least two, and usually four, transits each year, around the dates April 3 and October 6.

The new planet was given a name—*Vulcan* (French, Vulcain). This was the name Babinet used for one of his gaseous intramercurial masses invoked to explain the prominences seen during a total eclipse of the Sun. It was also the name of the Roman god of fire, fitting for a planet lying so close to the Sun. Just when Vulcan was first adopted for Lescarbault's planet is not entirely clear. Neither Lescarbault in his letter of December 22, 1859, nor Le Verrier in his presentation to the Académie des Sciences of January 2, 1860, used the name. On the other hand, the Abbé Moigno commented in an article in *Cosmos*, February 3, 1860:

> It will be more useful to make another observation of the planet than to argue about a name. The latter can well wait until the planet shows itself again. And yet names have been proposed by a great many of our correspondents, several of whom, at their own initiative and without our encouragement, favour Vulcan—the only name, they say, which could possibly be considered for an intramercurial planet. Indeed, most probably it will be called Vulcan. M. Babinet, by the way, first gave the name Vulcan long ago to the cosmic clouds by which he explained the rosy prominences observed at some solar eclipses.[19]

In the next issue of the same journal, published on February 10, J. C. R. Radau referred to Vulcan as the definite name of the planet.

Radau, a professor of astronomy at Königsberg, recalculated the orbital elements from Lescarbault's observations. Assuming a circular orbit, he put the mean distance at 0.143 AU, from which followed a period of revolution of 19.7 days. But the orbit might be highly eccentric; indeed, he noted, this was even probable. In that case, other observations would be necessary to define its orbit precisely.

Perhaps the planet *had* been seen on earlier occasions. The archives were thoroughly ransacked for such data, just as they were following the discoveries of Uranus and Neptune. Even before Lescarbault's observation was published, Rudolf Wolf had, as mentioned earlier, sent Le Verrier a list of suspicious spots observed on the Sun. And within days of Le Verrier's presentation to the Académie des Sciences, a curious letter appeared in the London *Times*, rather ominously recalling the international furor over Neptune. "Although I should deeply regret," wrote Benjamin Scott, a Fellow of the Royal Astronomical Society and Chamberlain of the City of London, alluding to his hitherto secret observation, "that any circumstances should interrupt the present amicable relations between France and Great Britain . . . yet in justice to my country's honour, I must enter an appearance for her of prior discovery in this case. The body alluded to was first seen by Mr. Lofft (an Englishman, I believe) on the 6th of January 1818, and by myself at or about Midsummer in 1847; and, whenever since that date I have publicly lectured on the solar system, I have always asserted, as a fact known to myself, that a third inferior planet existed in our system."[20] The circumstances of Scott's observation were somewhat dubious; after a lapse of 23 years, he could no longer even remember the exact date of the observation, which consisted of no more than the glimpse of a dark spot on the Sun made near sunset. It was confirmed at the time by Scott's son, then only 5 years old, who commented, perhaps with rather more insight than his father, "I see a little balloon on the Sun!" The spot or balloon was no longer present the following morning. Scott passed the information on to Richard Abbatt, another Fellow of the Royal Astronomical Society, who advised that he had probably seen an ordinary sunspot, and nothing more was done to follow up the report. Scott now regretted, "I had not sufficient self-reliance to induce me to make a public announcement of this remarkable observation, although I never lost faith in what I had seen."[21]

The observation by Capel Lofft, an amateur astronomer at Ipswich, was published in the *Monthly Magazine* for March 1, 1818. Lofft, using several small telescopes owned by himself and his friends Acton and Crickmore, had recorded an apparent interloper on the Sun on January 6. "When I first saw it," he recalled, "[it was] somewhat about one-third from the eastern limb; subelliptic, small, uniformly opaque." About 3½ hours later, it "appeared to Mr. Acton considerably advanced, and a little west of the Sun's centre." It had not been visible on January 4 or 8; it was not visible just before sunset to Crickmore on January 6, which induced Lofft to remark that "its progress over the Sun's disc seems to have exceeded that of Venus in transit." Lofft himself speculated it might have been a comet in transit across the Sun.[22]

That seemed reasonable as there was at least one other case where this was believed to have happened—the head of the comet of 1811 had actually passed in front of the Sun's disk, on June 26, 1811. Olbers calculated the event a month after it had actually taken place, but as soon as he published his result, "several observers . . . brought forward accounts of singular spots perceived by them upon the Sun at the time of the transit, and the original drawing of one of them, by Pastorff of Buchholtz, has been preserved."[23] J. W. Pastorff was one of the most prolific observers of fast-moving spots upon the Sun. His drawing of June 26, 1811, does indeed show a "round nebulous object with a *bright* spot in the center, of decidedly cometary aspect, and not in the least like an ordinary solar 'macula.'[24] It was later shown by J. R. Hind that this spot was not in the right position to be the comet, moreover comets are such small, flimsy bodies that they could never be seen in this manner. Whatever Pastorff may have seen—if indeed he saw anything at all—it could not have been the head of the comet of 1811.

Other suspicious observations soon surfaced; Wolf enlarged his list of unusual sunspots, and further entries were added by the English solar observer Richard Carrington, who from his private observatory at Redhill, Surrey, carried out his own series of sunspot observations since 1853.[25] Others, notably C. Haase and Rev. T. W. Webb, added still other observations to the list. The more widely discussed entries were:

1762 November 19. Lichtenberg saw, with the naked eye, a great round spot of about 1/12 the diameter of the Sun, traverse a chord of 70° in 3 hours.

1764. Between 1 and 5 May. Hoffmann near Gotha saw, with the naked eye, a large round spot, of about 1/15 the diameter of the Sun, traverse it slowly from north to south.

1798 January 18. D'Angos, at Tarbes, saw a slightly elliptical, sharply defined spot, about halfway between the center and edge of the Sun, which passed off about 25 minutes afterwards.

1802 October 10. Fritsch, at Magdeburg, saw a spot moving 2 minutes of arc in 3 minutes of time; not seen after a cloudy interval of 4 hours.

1818 January 6. Lofft's observation described above.

1819 October 9. Stark, canon of Augsburg, saw a well-defined round spot, about the size of Mercury, not to be seen the same evening.

1820 February 12. A circular well-defined spot, with a circular atmosphere and orange-gold tint, not seen again the same evening, is said to have been seen by two independent observers, Stark and Steinhübel. It crossed the Sun in about 5 hours.

1839 October 2. Decuppis, at the Roman College, saw a perfectly round and defined spot, moving at such a rate that it would cross the Sun in about 6 hours.

1847. Scott's observation described above.

1849 March 12. G. C. Lowe and J. Sidebotham watched for half an hour a small round black spot traversing the Sun. Sidebotham later wrote: "We were trying the mounting and adjustments of a 7-inch reflector we had been making. . . . At first we thought this small black spot was upon the eye-piece, but soon found it was on the Sun's disc. . . . The only note in my diary is the fact of the spots being seen—no time is mentioned, but if I remember rightly it was about 4 o'clock in the afternoon."[26]

1858 August 1. Wilson, at Manchester, observed a perfectly round spot and noted: "By watching it closely I fancied I could really see it moving, and leaving the cluster of [ordinary] sunspots on its right."

A number of these observations were suspiciously vague in the details, lacking exact dates or times of transits. And yet the powers of suggestion were strongly at work. A heavy dose of wishful, or at least uncritical, thinking gave Vulcan a somewhat more substantial existence, and it grew, briefly, into a compelling mirage.

The critical floodgates had burst open. After Lescarbault's discovery, Vulcan-mania swept like wildfire. Ironically, the source of this astronomical hysteria was no mesmerist or crank but the Imperial Astronomer of the Paris Observatory, the Giant of Science, he of the austere formulae and meticulous analysis who after overcoming his initial doubts had fallen hopelessly in love with Lescarbault's planet. Vulcan became for him, indeed, an idée fixe. His attitude was explained only by his profound commitment to Newton's gravitational theory and his deep psychological need for such a planet.

Strangely, Vulcan did not even solve the problem for which its existence had been invoked—that of providing the missing mass needed to produce the anomalous advance of Mercury's perihelion. Le Verrier himself, in a postscript to the paper in which he first disclosed Lescarbault's observation, revealed the mass of the new planet (based on the diameter Lescarbault had reported) could only be 1/17 that of Mercury. Thus, it would take 20 Vulcan-sized planets to account for the anomalous advance of Mercury's perihelion. Le Verrier was apparently unconcerned; he thought that Vulcan was probably only the largest of a number of asteroidal bodies making up a ring. Consistent with this notion, the various observations Wolf had collected seemed to indicate other intramercurial planets had been seen from time to time. A rising young American celestial mechanician, Simon Newcomb, objected that the existence of such a ring ought to introduce other complications to the motion of Mercury, such as an observable perturbation of its nodes.[27] His arguments proved to be flawed, however, as he himself later admitted.[28] In any case they attracted scant attention. At the moment, the need for other small planets and their possible dynamical implications were overshadowed by interest in Vulcan.

eleven

The Phantom
of an Anomaly

Wolf culled from his observations three (January 18, 1798, October 10, 1802, and October 19, 1819) which seemed to be reconcilable with Lescarbault's planet, and passed the information on to Le Verrier.[1] Radau, however, objected that the planet, or rather black spot, of 1798 could not be Vulcan if it was to be identified with the planet of 1802 and 1819. Using the latter observations only in combination with that of Lescarbault, Radau deduced the period of Vulcan to be 38.5 days, not 19.7 days, and declared the planet lay in an orbit inclined to the ecliptic by less than 1° 5'. This was rough, to say the least, but it did lead to definite expectations: Transits of Vulcan across the Sun were expected to occur on March 29, and April 2, 4, and 7, 1860.[2] Astronomers around the world awaited these events but though the surface of the Sun was carefully scrutinized the result was . . . nothing.[3]

Vulcan's failure to materialize was embarrassing. Even more so was a paper, "Sur la Nouvelle Planète announcé par M. Lescarbault" (On the New Planet announced by M. Lescarbault), written by Emmanuel Liais, another French astronomer. Liais was a known rival of Le Verrier and a close friend of Alexis and Eugène Bouvard, who were staff members of the Paris Observatory. At the time of Lescarbault's observation, Liais was employed with the Brazilian Coastal Survey. He was known as a skilled observer, noted, among other things, for his independent discovery of the Great Comet of 1861 (within a day of John Tebbutt in New South Wales). Liais wrote his article on Vulcan on March 8, and it appeared in the leading German journal, *Astronomische Nachrichten*, a week after the last of Radau's predicted transits had passed unwitnessed. His paper proved to be a bombshell—a decisive turning point in the Vulcan affair.[4]

Even before he published Liais had distinguished himself as among the most aggressive members of the "happy coincidence" school regarding the discovery of Neptune. Of Neptune he had declared: "It was certainly near the position given by Le Verrier; but the orbital elements were different, and the agreement in position was due to sheer chance. At the time, of course, there was great praise for Le Verrier; but time has passed since this first enthusiasm. . . . In fact Le Verrier's hypothetical planet, near the position where Neptune was found, does not exist, and must be relegated to the realm of fiction." These were some of his milder comments! He also said: "Le Verrier took the question as being a simple calculation, and did not rise to considerations about the real nature of the problem—as Arago, with his touch of genius, would certainly have done." And further: "To Galle, therefore, and not to Le Verrier, the honor of the discovery, as to Newton, and not to the apple, that of universal gravitation."[5]

Liais said he first learned of Lescarbault's observations, in apparent confirmation of Le Verrier's expectation of an intramercurial planet, only in February 1860. By then, he noted ironically the supposed discovery had already been rather too hastily endorsed by the Académie des Sciences:

> Whoever concerns himself seriously with physical astronomy knows a thousand different causes of such illusions, especially with glasses in cardboard. But M. Le Verrier believed only in his calculations. He had no doubt whatsoever. The planet had been found. True, circumstances included some details hardly typical of an orthodox astronomer; but everything was quickly arranged, and the affair passed the *Institut* without objection.[6]

Since Liais himself had observed the Sun frequently from San Domingos between January and July 1859 (the specific object of his investigation being brightness comparisons of different parts of the solar disk), he went back and consulted his notes. He found what he was looking for. He had been carrying out a careful examination of the Sun's surface at the exact instant Lescarbault had supposedly observed the planet's transit, but he had seen . . . precisely nothing.

The details are as follows. On March 26, 1859, between 11:04 and 11:20 a.m., Liais made a series of observations comparing the relative intensities of the center and limbs of the Sun. Then, between 12:42 and 1:17 p.m., he made another series comparing the poles with the equator (the break between the two series was due to clouds, he explained). Since the longitude of Orgères differed from that of San Domingos by 3 hours, the entry of Lescarbault's planet onto the Sun should have taken place at 1:05 p.m. at San Domingos, its exit at 2:23. At 1:17, when Liais stopped observing, the planet should have been on the Sun's disk for 12 minutes. And yet of that part of the disk all he noted was: "Region of very uniform intensity, consisting of tiny specks."

"Well," he concluded, "is it plausible that in an investigation of the physical constitution of the Sun and using a magnification twice that of M. Lescarbault I should not have noticed a sunspot 79° from the equator, when for each comparison I carefully examined this area for apparent variations in intensity? To the contrary, I am in a position to deny, and deny positively and absolutely, the passage of a planet across the Sun at the time indicated."[7]

Liais went so far as to impugn the veracity of the doctor of Orgères, attacking, through him, Le Verrier himself: "Lescarbault contradicts himself," he wrote, "in first asserting that he saw the planet enter upon the Sun's disk, and then admitting to Le Verrier that it had been on the disk some seconds before he saw it and that he had merely inferred the time of its entry from the rate of its motion afterwards. But if this one assertion be fabricated, the whole may be so."[8] The charge of dishonesty effectively precluded a reply; Lescarbault was silent, and the Abbé Moigno remarked the accusation was of such a nature as to excuse him from any obligation to refute it. ("This was an error of judgement, I cannot but think," wrote Richard Proctor, "if an effective reply was really available.")[9] Le Verrier, too, attempted no response to Liais's paper.

Certainly the note of rivalry and bitterness Liais injected into his paper damaged his case. But he had seriously damaged Le Verrier's hypothesis. According to Liais, at least one prominent European astronomer (not, unfortunately, identified) congratulated him "for opposing the fantasies regarding the existence of a planet between Mercury and the Sun."[10]

Liais had other arguments that cast doubt on the existence of Vulcan. Many observations of unusual spots on the Sun were illusions, he suggested. In some instances, they were due to subjective effects such as eye fatigue, in still others to defective optics. Small dark specks, evanescent in nature, frequently appeared on the Sun, and if the observer was careless he might become confused and receive the impression of a single moving spot. When the Sun was low, cloud bands sometimes cut across the disk of the Sun; they were sometimes associated with dark detached points, such as he himself had noted in the Bay of Rio de Janiero on September 8, 1859. Any of these effects could give the impression of bodies crossing the solar disk.[11]

Finally, just because Lescarbault saw a dark body on the Sun did not prove the existence of an intramercurial planet. It was impossible from Lescarbault's observations to pinpoint the object's exact location, apart from its lying somewhere on a line between the Earth and Sun. Rather than an intramercurial planet, it was much more likely Lescarbault had encountered a small body in the neighborhood of the Earth. Such a body, moving obliquely to the visual-ray, would show up on the solar disk, move off and disappear. Its rate of motion would appear similar to that of an intramercurial planet.[12]

Again, if Vulcan actually existed, Liais argued, it ought to be brighter than Le Verrier supposed. Lying closer to the Sun than Mercury, it would, despite its small size, shine with far greater refulgence. He estimated that, if one accepted Lescarbault's observations at face value, it would have something like seven times the brightness of Mercury. It might not be visible with the naked eye, owing to its proximity to the Sun, although even this was by no means certain since Liais himself, and many others for that matter, had seen Mercury when it was only 7° or 8° from the Sun. It would certainly be easily visible in a telescope. It also seemed absolutely incredible it should have escaped notice during total eclipses. Finally, Liais doubted the anomalous motion of the perihelion of Mercury itself, preferring to admit a possible error in the varying angle between the plane of the Earth's equator and its orbit, that is, the obliquity of the ecliptic, of about 2° and a consequent increase of the mass of Venus in order to explain the discrepancies in the transit observations.

Critical observations in quest of Vulcan were advocated by Faye, even before Lescarbault's unexpected sighting had turned up, at the total eclipse visible from Spain in July 1860. Shortly before the eclipse, the noted British scientist David Brewster published an article on Vulcan in the *North British Review*. Brewster was generally sympathetic to Le Verrier's hypothesis, but concluded with a balanced assessment:

> If, after the severe scrutiny which the Sun and its vicinity will undergo before, and after, and during his total eclipse . . . no planet shall be seen . . . we will not dare to assert that it does not exist. We cannot doubt the honesty of M. Lescarbault; and we can hardly believe that he was mistaken. No solar spot, no floating scoria, could maintain, in its passage over the Sun, a circular and uniform shape; and we are confident that no other hypothesis but that of an intramercurial planet can explain the phenomena seen and measured by M. Lescarbault—a man of high character, possessing excellent instruments, and in every way competent to use them well, and to describe clearly and correctly the results of his observations.

And yet:

> Time . . . tries facts as well as speculations. The phenomenon observed by the French astronomer may never be again seen, and the disturbance of Mercury which rendered it probable, may be otherwise explained. Should this be the case, we must refer the round spot on the Sun to some of those illusions of the eye or of the brain, which have sometimes disturbed the tranquility of science.[13]

The Spanish eclipse came and went. Vulcan was searched for, but as at the transits expected in March and April, it did not appear. Now an attitude of skepticism prevailed.

Lescarbault's sighting and the irregular collection of doubtful reports assembled by Wolf and others could hardly offset the fact the planet had not been seen by some of the greatest and most experienced solar observers, such as Schwabe and Carrington. As William Frederick Denning, an English amateur, later summed up:

> Through all the years during which these experienced observers scanned the Sun with devoted pertinacity, and amongst all the well-nigh innumerable host of sunspots which they examined, not a single instance can be found in which the records distinctly refer to a planetary body in transit, and this cannot fail to be regarded as a fact tending to negative, in the strongest manner, the isolated descriptions which have been adduced upholding the theory of a new planet.[14]

And yet ignominy was averted. Vulcan, or what was taken for Vulcan, unexpectedly appeared again on March 20, 1862. The circumstances were strikingly similar to those of Lescarbault's observation. W. Lummis, of the Manchester, Sheffield, and Lincolnshire Railway Company's Office at Manchester, communicated the information to J. R. Hind, then Superintendent of the *Nautical Almanac*. Observing the Sun with a 2 ¾-inch telescope, Lummis claimed to have seen "a small black spot more regular and better defined than usual." He kept the object in view for about 20 minutes, and showed it to a colleague; but before it had exited the solar disk, he was called away by his official business. Lummis furnished Hind with the following additional details:

> As to the positions of the spot I regret that I cannot give them with much more precision than shown in the rough sketch sent to you, which was taken from one made at the time I witnessed the transit. I had no instrument but the telescope, and I measured with a small strip of card the distances of the spot from the Sun's edge. From the time when I first observed it, 8h 28m a.m., (Manchester time) to 8h 50 m, the spot had moved over about 12' of arc, as nearly as I can judge by the eye, and its size or apparent diameter I should take to be about 7".... I had for several mornings been observing the Sun, which I noticed was particularly free from spots.[15]

Lummis's object was immediately hailed as Vulcan. Thus Hind noted:

> the period of revolution assigned by M. Le Verrier from the observations of 1859 was 19.70 days. Taking this as an approximate value of the true period, I find, if we suppose 57 revolutions to have been performed between the observations of M. Lescarbault and Mr. Lummis, there would result a period of 19.81 days. On comparing this value with the previous observations in March and October, when the same object might have transited the Sun at the opposite node, it is found to lead to October 9, 1819 . . . and on this very day Canon Stark has recorded [a similar] observation.[16]

Valz, comparing Lummis's observation with that of Lescarbault, arrived at a period of 17.5 days for Vulcan, while Radau found 19.9 days. Le Verrier was ecstatic.

However, Lummis's document was hardly of great precision. The observer had been taken completely unaware, and as with the original sighting by Lescarbault, doubts began to creep in. Although Le Verrier continued to regard the observation as a triumphant confirmation of Lescarbault's planet, C. H. F. Peters, of Hamilton College, New York, announced he himself had observed Lummis's object. Another seasoned observer of sunspots, Gustav Spörer in Germany, had also observed it. Their records showed it corresponded to nothing more than two ordinary sunspots. Evidently Lummis had seen one and then, 20 minutes later, the other. Confusing them, he concluded he was seeing a single spot with rapid motion. The positions of the two spots were such as to have readily originated the mistake, and agreed in all details with the descriptions given by Lummis. Another embarrassment for Vulcan.

Like the phantom satellite of Venus of the 17th and 18th centuries, Vulcan managed to reappear just often enough to maintain a shadowy existence among true believers. Coumbary, at Constantinople, on May 8, 1865, saw a small object crossing the Sun, which detached itself from a group of sunspots.[17] At the total eclipse of August 7, 1869, a "little brilliant" was seen just outside the corona by four observers, all inexperienced amateurs, at St. Paul's Junction, Iowa.[18] At the same eclipse, Simon Newcomb of the U.S. Naval Observatory carried out a search from his station near the Court House in Des Moines. Just before the eclipse, he asked C. H. F. Peters to join him, but Peters declined; he had come to Des Moines "to observe the eclipse," he said, "not to go on a wild goose chase for Le Verrier's mythical birds."[19] In the event, Newcomb saw nothing unusual. Another distinguished American astronomer, Benjamin Apthorp Gould, Jr., noted in his report he had seen the star 82 Cancri but had not encountered any other stellar body. Since this star was at the time in the same direction as the "little brilliant," he simply assumed it was the object remarked at St. Paul's Junction.[20] But uncertainties remained. Hind, emphasizing 82 Cancri is of only 6th magnitude, suggested it might well have been the other way around, that Gould had mistaken an intramercurial planet for the star.[21] Parenthetically, the star fields of Cancer would figure once again in the Vulcan affair—at the eclipse of 1878.

Denning, in Bristol, England, organized a massive search for Vulcan, involving no less than 16 observers stationed at various points throughout England, during the time of the planet's expected passage near its spring node, March–April 1869. The result was as so often before—nothing. An even larger effort the following year led to the same negative result.[22] Hind, long a supporter of the Vulcan hypothesis, carried out a complete reexamination of the whole question in 1872 and concluded it was probable the planet would transit the Sun on March 24, 1872. Nothing was reported in Europe, America, or Australia, but a lone observer in Shanghai telegraphed that the "predicted circular spot" had been seen at 9 a.m. on that date. There was no reference to its motion. Presumably it was an ordinary sunspot, but in any case no additional details were forthcoming.[23]

Throughout the controversy Le Verrier remained in the background; his faith in the planet's existence never wavered. He took pains to publish every possible sighting in the *Comptes rendus* of the Paris Académie des Sciences. But there was other work to do. Once more, he was immersed in a massive project of planetary calculations.

After completing his great memoir on the motion of Mercury in 1859, 2 years later he published similar treatises on the motions of Venus and Mars. In 1868, he was awarded the gold medal of the Royal Astronomical Society for his memoirs on the Sun and inner planets. Obsessed as always with his great dream of extending ever farther the reign of Newtonian law, he was by then totally absorbed in the theories of the giant planets: Jupiter, Saturn, Uranus, and Neptune. These efforts were praised by John Couch Adams, who was certainly in a position to judge. "The theories of the mutual disturbances of the larger planets," he declared, "are far longer and more complicated than those of the smaller, so that all that M. Le Verrier had yet done might be almost regarded as merely a prelude to what still remained to be done. Increased difficulties, however, far from deterring, seemed rather to stimulate him to greater exertions."[24]

The notion a prophet is not without honor except in his own country applied, to some extent, to Le Verrier. Among foreign scientists, he was regarded as the Giant of Science. Perhaps a bit stiff in personal manner, he was generally regarded as amiable enough. In appearance he was still, at 60, handsome: "a decided blond, with light chestnut hair turning gray, slender form, shaven face, rather pale and thin, but very attractive, and extremely intellectual features."[25] It was his colleagues, his own countrymen, who found him insufferable.

His well known violent temper had become worse with age. One colleague referred to him as a *mauvais coucheur* (a bad bedfellow);[26] yet another told the young English scientist J. J. Thomson who wished to call on him during a visit to Paris, "I do not know whether M. Le Verrier is actually the most detestable man in France—but I am quite certain that he is the most detested."[27] Apparently Le Verrier adopted the same lionlike demeanor to his subordinates at the Paris Observatory he had initially shown to the cringing Lescarbault.

In 1862, he summoned Camille Flammarion into his office. Flammarion had just published his book, *La Pluralité des mondes habités,* and woke to find himself famous. However, rather than congratulate his young assistant as might be expected, Le Verrier dismissed him: "I see, Monsieur, that you no longer need remain here; no, you may resign."[28] Although later hired back, Flammarion remained hostile to Le Verrier. With Liais he became one of the sharpest critics of Vulcan. After Le Verrier's death, Flammarion was able to bring himself to admit that Le Verrier's employments had been "noble," but even then could not resist qualifying his praise with the comment that they "might have been still more useful if he had possessed a more sociable character and a more disinterested love for the general progress."[29]

Things came to a head in 1870. Despite his growing unpopularity among his colleagues, Le Verrier's unrivaled scientific achievements and the support of Napoleon III—the imperial astronomer had always been a monarchist and imperialist—had until then placed him in a virtually unassailable position. Ministers were afraid to move in the matter, and because he was a man of such violent temper everybody kept out of his way. They tried to keep him in check by appointing a Board of Advisers, without whose consent and recommendation nothing could be done. There was not a member of the board with whom Le Verrier was on speaking terms, and for 6 months he refused to have anything to do with this body. At last, finding this would not do, he reluctantly agreed to attend a meeting; it is said to have ended with three of the advisers kicking Le Verrier out of the room. According to a contemporary report, "all of the meetings he attended ended in a row; the scientific men began to be very tired of such things, and all of the assistants employed in the observatory sent in their resignations."[30] After this open rebellion against his rule, Le Verrier was finally forced to resign.

His place was taken by his long-time rival Charles Delaunay. Although he had not made a sensational discovery like that of Neptune, Delaunay was an equally accomplished master of celestial mechanics. His investigation of the Moon's motion was, as Simon Newcomb later noted, "one of the most extraordinary pieces of mathematical work ever turned out by a single person. It fills two quarto volumes, and the reader who attempts to go to through any part of the calculations will wonder how one man could do the work in a lifetime."[31] In personal manner, Delaunay was as unlike

Le Verrier as he could possibly be; according to Newcomb, "one of the most kindly and attractive men I ever met."[32]

Soon after Le Verrier severed his connection with the observatory, war broke out between France and Prussia. Bismarck's Prussian army proved more than a match for Napoleon III's army; after sweeping across the French countryside, it formed a blockade around Paris—indeed, German artillery rumbled through Orgères itself. The blockade proved to be most effective. Food became so scarce the inhabitants ate animals auctioned off by the zoological gardens. Nevertheless, the Académie des Sciences continued to meet regularly (the legal quorum being three, this did not imply a large attendance!). Prudently, Le Verrier left the city before the siege began, but another French astronomer, Jules Janssen, was forced to make a romantic escape by balloon while attempting to observe the total eclipse of the Sun of December 22, 1870. He succeeded in getting to Oran, Algeria, which lay within the path of totality, only to find himself hopelessly "shut behind a cloud-curtain more impervious than the Prussian lines."[33]

Paris at last surrendered to Bismarck in January 1871. The Republic had already been proclaimed the previous September. Napoleon III was captured at Sedan and deposed, and Adolphe Thiers was elected Chief of the Executive Power of the French Republic. Still the situation in Paris continued to deteriorate until in the spring, it exploded in a spate of civil violence—the so-called Paris Commune of 1871. On April 23, Newcomb, who was visiting Delaunay at the Paris Observatory, wrote to his wife of a deserted and threatened institution, and of the necessity of hoisting a red flag to appease the Communards.[34] Finally in May, after the brutal suppression of the Communards by Thiers and the army, Parisian affairs slowly returned to some semblance of normalcy.

Le Verrier returned to Paris, but his exile from the observatory lasted another year. It ended only with Delaunay's death. The latter had an extreme phobia of water, based apparently on the fact that both his father and brother had drowned. He was accustomed to taking every precaution to avoid entering a boat; in all his life he had only once crossed the English Channel—on the occasion when he received the gold medal from the Royal Astronomical Society. In autumn 1872, he was walking along the beach at Cherbourg, and was invited by some boatmen to sail with them; inexplicably he ignored his rule. Alas, once the boat had ventured some distance from land a squall blew up, and Delaunay and the rest of the party were drowned.

Le Verrier, who had been a close personal friend of Thiers, was now recalled to his former position as director of the Paris Observatory. He immediately resumed his labors on the giant planets and began catching up on a correspondence massively in arrears because of his absence. To one American astronomer, E. S. Holden, he apologized for not answering a letter for 2 years, pleading "the great political preoccupation of this epoch."[35] Paper followed paper in rapid succession. In 1873, he presented to the Académie des Sciences his complete theory of the motions of Jupiter and Saturn, a year later that of Uranus, to which he had returned for the first time since the investigation that had led to the discovery of Neptune, and then, finally, he made a start on the theory of Neptune itself. Tables for Jupiter appeared in 1874, for Saturn in 1875, and for Uranus in 1876, when he received for a second time the gold medal of the Royal Astronomical Society. "That any one man," noted Adams in his remarks on this occasion, "should have had the power required thus to traverse the entire solar system with a firm step, and to determine with the utmost accuracy the mutual disturbances of all the primary planets which appear to have any sensible influence on each other's motions, might well have appeared incredible if we had not seen it actually accomplished."[36]

These were massive efforts. And yet it is strange it was the motion of Mercury, the least of all these planetary members, that should in the end have proved the linchpin of the whole. After years of awakened hopes and disappointed predictions, Vulcan—part of the missing mass Le Verrier required to salvage Newtonian mechanics—remained singularly elusive.

Le Verrier's apparent interest in an intramercurial body (or bodies) had been dormant for several years as he had struggled to scale the summits of the dynamical interactions of the giant planets. However, he had given no hint of abandoning his earlier conclusions. In 1874, he flatly asserted: "There is, without doubt, in the neighborhood of Mercury, and between that planet and the Sun, matter hitherto unknown. Does it consist of one, or of several small planets, or of asteroids, or even cosmic dust? Theory alone cannot decide this point."[37]

However, the real reawakening took place in 1876. On April 4, a German astronomer, Heinrich Weber, telegraphed from Peckeloh, northeast China, to report he had seen a transit of a small round spot on the Sun. Unfortunately, he did not comment on its rate of progression. His telegraph reached Wolf at Zurich and Julius Schmidt in Athens. All three scanned the Sun closely the following day, but the spot could not be found. Had it, indeed, been the fugitive planet?

Le Verrier himself had no doubt. Wolf had informed him it seemed to be identical with Lescarbault's object—with good reason, since Le Verrier's initial calculations had indicated Vulcan ought to pass near one of its nodes on or about April 3 each year, and Weber's sighting had occurred on April 4.[38] Though on many similar occasions the planet failed to keep its appointments, Weber's report was accepted without hesitation, at least by those who already believed in Vulcan. Le Verrier was overjoyed; so was the Abbé Moigno, who forwarded congratulations to Lescarbault. Strangely, Lescarbault seems to have received the information with indifference. According to Richard Proctor, "he, it seems, has never forgiven the Germans for destroying his observatory and library during the invasion of France in 1870, and apparently would prefer that his planet should never be seen again rather than that a German astronomer should see it."[39]

According to Wolf, by assuming a period of 42.2 days, Weber's transit would have been the 148th such event since Lescarbault's observation of March 26, 1859.[40] Le Verrier set aside his work on the tables of Neptune to pore over historical sightings of possible intramercurial planets—over 20 between 1802 and 1876. He found he could divide them into four groups, each of which might represent a different planet. The strongest association seemed to be indicated by five observations: by Fritsch on October 10, 1802; Decuppis on October 2, 1839; Sidebotham on March 12, 1849; Lescarbault on March 26, 1859; and Lummis on March 20, 1862. (Hind, later, called to his attention yet another observation, which seemed conformable to this grouping, by Canon Stark on October 9, 1819.)

On September 12, 1876, Le Verrier communicated with Wolf to suggest that instead of 42.2 days, a period of 28.01 days might better fit the observations.[41] Continuing to juggle, he wrote to Wolf again on September 21, suggesting if Canon Stark's observation of 1820, Lescarbault's of 1859, and Weber's of 1876 represented (as he thought) three apparitions of the same body, it ought to transit the Sun on October 2 or 3 if the period were 28 days or on October 9 or 10 if, as Wolf believed, the period were actually 42 days. At the moment Le Verrier was unable to choose between the alternatives. Nevertheless, he told Wolf, "I would prefer to postpone the discussion and observe both carefully. We will have leisure to discuss the results later."[42] He promptly alerted observers around the world to watch for the expected transits.

Still, matters remained confused. Hind, for instance, opined that the Lescarbault and Lummis objects could not have been the same object at all. Le Verrier wrote in a quandary to Wolf:

> If these are not observations of the same body, what becomes of the period which you derived from them, which at the time seemed so entirely satisfactory? As for my own period of 28 days, it was derived from your results. . . . One thing at least we can agree about: we must continue to look assiduously between now and October 10, and then every year hereafter between the end of March and the first part of April, and again between the end of September and the first part of October.[43]

Astronomers did scan the Sun October 2–3, but to no avail; the planet failed to appear. (Among those who kept up a vigil on the Sun was the ever-skeptical C. H. F. Peters, who was reported in the *Utica Morning Herald* to have done so "as a matter of courtesy, rather than of belief. . . . Diagrams were made and the positions of all spots visible at different hours accurately recorded. Nothing in the shape of a planet was discovered, as Dr. Peters was confident would be the result."[44])

The next day, October 4, there was yet another embarrassment. Weber's observation, which had created so much of the latest stir about Vulcan, had been completely discredited. His spot had been independently recorded and identified, it turned out, 5 hours before he had seen it in China, by Señor Ventosa at Madrid, who had recognized it as an ordinary sunspot though one without a penumbra. Any remaining doubts were removed when the Astronomer Royal, George (now Sir George) Airy telegraphed to astronomers the results of an examination of simultaneous photographs of the Sun taken at the Greenwich Observatory. "There was a small round spot," Airy announced, "in a group of faculae near the north-east limb in the place indicated by Herr Weber's observation. . . . There can be no question that the spot on the Greenwich photographs, which is the same as that observed by M. Ventosa, is an ordinary sun-spot without penumbra, and not an intramercurial planet."[45] The sunspot so recorded was not even completely round. "It is clear," Richard Proctor later admonished, "that had not Weber's spot been fortunately seen at Madrid and photographed at Greenwich, his observation would have been added to the list of recorded apparitions of Vulcan in transit. . . . I think, indeed, for my own part, that the good fortune was Weber's. Had it so chanced that thick weather in Madrid and at Greenwich had destroyed the evidence to show that what Weber described he really saw, although it was not what he thought, some of the more suspicious would have questioned whether . . . "the round spot on the sun" was not due to "one of those illusions of the eye or of the brain which have sometimes disturbed the tranquility of science."[46]

Le Verrier conceded the point, but maintained the reality of Vulcan. He now awaited the verdict of October 9–10, 1876. As so often, the disk of the Sun was intensively watched for the expected intruder. Again the planet preferred discourtesy and failed to appear.

For all the disappointments, the excitement did not die down immediately. Weber's spurious observation and Le Verrier's resurgence of interest had triggered another round of Vulcan-mania, and for a few months, reports broke out like measles. It was predictable enough. What amateur astronomer or untried novice did not feel the temptation to step forward with his own half-fledged observation of what the Abbé Moigno had called "the planet of romance," hoping to achieve the recognition he craved? Every small sunspot was remembered as the sought-for planet. Inevitably there were even a few cases of deliberate imposture. When it is considered Lewis Swift, charged in the 1880s with judging the merit of claimants to a cash prize for the discovery of new comets, confided that he received some 70 letters a day from people claiming they were entitled to the prize, "some containing the biggest kind of lies,"[47] it is surprising that reports of Vulcan were not more numerous than they were.

The *Scientific American* for October 21, 1876, published a letter from an amateur astronomer of Montclair, New Jersey, who claimed to have seen a transit of the planet on the previous July 23. In the same issue a Reverend E. R. Craven recalled Professor Joseph S. Hubbard had assured him repeatedly he had seen Vulcan with the Yale College telescope; unfortunately, "the transit was an entire surprise, and hence no notes were taken," and moreover since Professor Hubbard himself was by now dead, he personally could provide no further pertinent information (for instance, as to whether or not he had been sober at the time).[48] In a later issue of the *Scientific American*, W. G. Wright of San Bernardino, California, reported how he had taken inspiration from these correspondents and decided to take his own telescope and look at the Sun. At once he discovered the planet, moving swiftly north to south across the Sun, and acquired local celebrity status. John H. Tice of Louisville, Kentucky, wrote to announce he had seen an object twice transiting the disk of the Sun in September 1859 and again in June 1876. It was pointed out his object could not be identical with the Vulcan of Le Verrier and Lescarbault, since it was larger than the planet Uranus; Tice remained unperturbed. After all, Le Verrier had said that Vulcan might be only one of several intramercurial planets![49] By then, the *Scientific American* had had enough, and closed the correspondence.

twelve

Le Roi Est Mort

Another year thus passed without Vulcan any closer to being confirmed. Eighteen seventy-seven dawned. Once more the world turned its attention to the Paris Observatory. Within its formidable redoubt sat the great mathematician, meditating his stratagems. His resourcefulness had been sorely tested, but he would not be found wanting. Possessed of unusual stubbornness ("increased difficulties . . . far from deterring, seemed rather to stimulate him to greater exertions," as Adams had said), he once more revised his calculations. He decided Vulcan's failure to appear October 9–10, 1876, was because it moved in a highly eccentric orbit with an extreme inclination he put at 10.9°. He derived a new period: 33 days with the next transit due to occur on March 22, 1877, then not until October 15, 1882.[1] The prospects on the former occasion were uncertain, since his calculations showed the planet's trajectory would be almost tangent to the limb of the Sun. Clearly, its position could not be calculated with great accuracy.

All of this sleight-of-hand was a far cry from the supremely confident prediction that astounded the world in 1846 when the master declared, as Airy put it, "Look in the place which I have indicated, and you will see the planet well."[2] Now, however, the world was confronted by a series of increasingly erratic predictions; Le Verrier may have had no doubts about Vulcan's existence, but to others the planet, so often crying wolf, seemed to have no more corporeality than a ghost.

The *Scientific American* editorialized Le Verrier's views on the subject, which

> appear to be in a transition state, and our French mails each week bring us new statements from him, which of late have invariably failed to accord; in fact they often wholly differ from those enunciated seven days before. The reader will therefore understand that the data we now give ... represent merely stages of progress in M. Le Verrier's investigation, through which we are endeavoring to follow him. The latest dictum of the eminent astronomer is more logical than some previous announcements, but at the same time seems to contradict flatly his previous results. In lieu of Vulcan swinging in a regular orbit in equal periods about the Sun, we are now told that its orbit is highly eccentric, and that the planet behaves like Venus, making two transits within a few years, and then not repeating the passage for a century. This, of course, puts a stop to any ... off-hand calculation of future transits.[3]

To this pass had the master mathematician come. "The 'hidden parameter' hypothesis of Le Verrier, so compelling with Uranus, ha[d] now lost its attractions," wrote Norwood Russell Hanson.[4] The prospect long unthinkable of somehow tweaking the law of gravitation itself, modifying the foundations of Newtonian theory, became palatable again. "The possibility of rejecting the whole theory—replacing it with new foundations—becomes genuinely a live option," Hanson added. But not for Le Verrier. Every bit of his work showed he could not do this. Everything else would have been challenged and hunted to preserve the integrity of Newtonian theory.[5] In the end, he was reduced to clutching at straws. The real irony of Le Verrier's quest for intramercurial planets lay in the fact that none of the bodies so eagerly pursued, none of the doubtful specks against the Sun asserted to be of intramercurial origin could save Newtonian mechanics from Mercury's ruthless advance, which remained unresolved even assuming a raft of questionable data.

Nevertheless, Le Verrier persevered. He sent out the alert for the March 22, 1877, transit, and the world responded as best it could. Airy resigned himself to humoring the man whose request for a planet search he had failed to oblige once before—and missed Neptune. He sent telegrams to India, Australia, and New Zealand and asked for observations to be made every 2 hours, more frequently if possible, on the critical date. "Without saying positively that he believed or disbelieved in the existence of the planet, Sir G. Airy thought, since M. Le Verrier was so confident, that the opportunity ought not to be neglected by anybody who professed to take an interest in the progress of planetary astronomy."[6] Nor was his request neglected; observers around the world took up their posts. The Sun underwent a round-the-clock vigil, visual and photographic, throughout the 72 hours of March 21, 22, and 23. Again the massive search registered a blank. The elusive Vulcan once more escaped the nets of the planet trawlers. Perhaps, as Le Verrier had hinted, it had followed a trajectory just beyond the limb of the Sun and the expected transit had proved a near miss. Another straw to grasp at. Le Verrier's last chance to snare Vulcan had passed into history.

By now he was seriously ill and too weak to attend the weekly meetings of the Académie des Sciences. He was also absent from his desk in the Paris Observatory. His colleagues hoped his health had only been temporarily shaken by his arduous labors. A rest would revive him, and so it seemed to do, briefly. In early August 1877 he rallied enough to return to his duties. A telegram informed him of a startling discovery—not of the long-awaited Vulcan, but of two small satellites of Mars, found by Asaph Hall using the 26-inch Clark refractor of the U.S. Naval Observatory at Washington, D.C.

Le Verrier's recovery proved illusive. Liver cancer was eating away at him. There was no time left to make a fresh start on the problem of Mercury, even were he minded to do so—and there is no evidence that he ever was. In mid-September he was seized with the most violent symptoms, yet he lived to see the proof sheets of his tables of Neptune completed by his colleague, Jean Baptiste Aimable Gaillot. Death came September 23—the anniversary of his greatest triumph: the optical discovery of Neptune, at Berlin, by Galle so many years before. Shy of the camera but now too feeble to protest, he allowed a portrait to be taken on the very day of his death.[7]

"His life," noted his longtime colleague Villarceau, "was dedicated almost entirely to astronomical science; and posterity will rank his name with those of Newton and Laplace. He will be forever remembered for his masterful contributions to celestial mechanics. . . . His wife and children will have the consolation of knowing that the memory of the one they have lost lives on in the observatory. Adieu, great astronomer! adieu!"[8]

So Le Verrier took up his well-earned rest among other illustrious Parisians who sleep in Mt. Parnasse's cemetery. No more dreams of unknown worlds. But Vulcan does not expire with him.

"LE ROI
LE VERRIER,
L'ÉTAIT MORT"

thirteen

The Ghost
Goes West

Vulcan will always be associated first and foremost with Le Verrier's name, and rightly so, since it was his calculations that first lent credence to its existence. And yet it was an American astronomer, James Craig Watson (Figure 16), who became the leading figure in the strangest episode of this bizarre and eventful history. For Watson, the intramercurial quest was nothing less than an obsession.

Watson was born on a small farm near Fingal, 5 miles southwest of St. Thomas, Ontario, on January 28, 1838. His ancestors had come from Ireland in the mid-18th century, and settled in Pennsylvania.[1] In 1811, his grandfather pushed 700 miles further west, resettling the family in the (at the time almost unbroken) forests of Upper Canada. However, the farmstead he established provided at best a scanty existence. Watson's father, William, after trying successively to scrape out a living as a farmer, carpenter, and schoolmaster, gave up and resolved to seek his fortune elsewhere.[2]

At first the family planned to settle in Detroit. However, when Watson's mother, Rebecca, learned that Ann Arbor was only 38 miles further west along the same railway line, she decided they should move there. When they arrived, they were penniless.

16. James Craig Watson (1838–1880). (Courtesy of the Mary Lea Shane Archives of the Lick Observatory.)

"The following years were years of bitter poverty and want, and they left their imprint deep upon the afterlife of James Watson," his student George Comstock later wrote.[3] Wonderfully gifted and full of ambition, Watson could not be kept down. Once when his father was working as an assessor in Canada, he had filled in for him during an illness, and "made all the computations and made up the assessment rolls of the district without error."[4] He was only 9. Later, in Ann Arbor, he got a job at the same factory where his father was employed. On pointing out to his employer that the man in charge of the steam-engine was incompetent, he was assigned to run it himself, and did so with entire success. When the factory closed, he was forced to peddle books and apples at the railway station, but in his spare moments taught himself Latin and Greek. In due course Watson "worked successively at nearly all the trades that were practiced in the village."[5] He won particular notice for his ability as a machinist.

At 15, he matriculated at the University of Michigan, which offered a free education, and excelled in the classical languages he had practiced so diligently. Later he began to attend the lectures of Franz Brunnöw, a German astronomer. Trained under Encke at Berlin and a onetime associate of Benzenberg at Bilk, Brunnöw had been enticed to Ann Arbor to introduce German educational methods into the curriculum and to direct its new observatory. He was a skillful computer of orbits, the man who first announced to J. R. Hind and thence to the English-speaking world Galle's discovery of Neptune. He spoke in broken English and was by no means an inspiring lecturer. Inevitably his classes began to dwindle until Watson alone remained; nevertheless, whenever he was reproached for the small size of his class, Brunnöw is said to have exclaimed proudly, "Yes, I have only one student, but that one is Watson!"

Dividing his attentions between working through Laplace's *Mécanique Céleste* and learning practical astronomy in the observatory, Watson still found time to exercise his mechanical bent in making a 4-inch refracting telescope. He ground and polished the object-glass with his own hands and briefly considered making telescopes for a living. He always had great self-confidence and the very highest expectations for himself, and once, in a college notebook, entered his name as "James Watson the Astronomer Royal."

After graduation, Watson was hired as a salaried assistant at the observatory. In June 1859, 3 months after Lescarbault observed Vulcan, Watson became professor of astronomy, gaining Brunnöw's place, who had resigned to take a position at the Dudley Observatory in Albany, New York. After only a year, Brunnöw returned to Ann Arbor, thereby displacing Watson to the chair of physics. With Brunnöw's final retirement in 1863, Watson resumed the professorship in astronomy and was also formally named to direct the observatory. By then he was making a name for himself as a regular contributor to astronomical journals—he had 15 papers published by the time he was 21—supplying observations of comets and asteroids and computing their orbits and ephemerides. He was a remarkably rapid and accurate calculator. Gifted with a phenomenal memory, he was able to carry several steps in his head without writing anything down and is said to have once completed, at a single sitting, the Herculean task of computing the elliptic orbit of a comet. "The possession of this skill," remarks Comstock, "perhaps acted injuriously upon the character of his scientific work, as it led him to give much of his time, as a paid computer, to work which others of inferior talent could have done equally well, though less rapidly."[6]

Soon after becoming Director of the University of Michigan Observatory at Ann Arbor, Watson began a massive project: the preparation of a series of charts of the stars near the ecliptic. The value of such charts derived chiefly from the fact that, from a thorough survey of these stars, the existence of any intruder, asteroid, or comet could be recognized at once.

In 1863, the astronomical world was not yet jaded by the discovery of new asteroids. Watson himself had already tasted the excitement of such discovery. On October 20, 1857, he independently swept up asteroid (47) Aglaia, only to find a German astronomer, R. Luther, had found it several weeks before. Now he took up the quest in earnest. Assisted by his charts and his memory, which allowed him to recognize "every star ... in its place" and "every variation from the normal position, brilliancy, and tint,"[7] his efforts were duly crowned with success. On September 14, 1863, he captured (79) Eurynome—the first of more than a score of such finds.

Meanwhile Watson's project of preparing charts of the stars along the ecliptic was being pursued, independently and with no less diligence, by C. H. F. Peters (Figure 17). Later Watson's rival, indeed his nemesis, Peters was if anything an even more colorful character than Watson himself.

Christian Heinrich Friedrich Peters was born September 19, 1813, at Coldenbüttel, Schleswig (then part of Denmark). In 1836, he received his doctorate under Encke at the University of Berlin.

17. Christian Heinrich Friedrich Peters (1813–1890). (Courtesy of the Mary Lea Shane Archives of the Lick Observatory.)

Failing in a bid to get a position at the Copenhagen Observatory, Peters went to Göttingen for further studies, and there befriended a young geologist, Sartorius von Walterhausen. The two young men set out for Sicily, and distinguished themselves by making a detailed study of the volcano Mt. Etna. Peters next accepted a position as head of the geodetic survey of Sicily. Part of his time was spent in Naples, where with a 3½-inch refractor he observed sunspots and discovered a new comet. When Sicily revolted against King Ferdinand II of Naples in 1848, Peters joined the revolt and was duly removed from his post. He managed to escape on board an English ship bound for Malta, then returned to Sicily to help fortify Catania and Messina. Finally, after Napoleon III's army overran the island in 1849, he fled, via France, to Constantinople in Turkey. When he arrived, all he had in his pocket was enough to buy breakfast or a cigar—he chose the cigar!

In Constantinople, the resourceful Peters learned to converse in Turkish and Arabic and became scientific adviser to Redshid Pasha, Grand Vizier to Sultan Abdul-Mejid II. The Sultan, a man of scholarly tastes, had acquired a fine 11-inch refractor from the celebrated Viennese optician, G. S. Plössel, and proposed Peters would be the man to put it to good use. "But Redshid Pasha's power and protection were not sufficient to overcome the antagonistic influences within the palace, nor could astronomical science, which would not stoop to rule the planets, prevail against the astrologers," according to a newspaper account of the time.[8] Peters was finally forced to give up his designs on using the telescope. In 1854, the Crimean War broke out, and Peters left Turkey for London, later emigrating to America, where he became a protégé of Benjamin Apthorp Gould, Jr., at the U.S. Coast Survey in Washington, D.C.

A year later, Gould was named director of the Dudley Observatory, and Peters followed him to Albany. Almost at once, however, he became involved in a bitter dispute between Gould and the trustees of the Dudley Observatory. Basically, Gould thought a scientist ought to run the observatory; the trustees, the local businessmen who had financed it, were equally determined to retain control. Peters was never afraid to speak his mind (it would not be the last time he found himself in the midst of litigation),[9] and in this particular matter he sided with the trustees. He compounded the problem by proposing to name the comet he discovered on July 25, 1857, Olcott's Comet, "after the very beloved and esteemed name of the distinguished citizen who is identified with the history of the erection of this observatory."[10] Inevitably, all this led to a complete break with Gould. Peters felt pressured to resign his position at the U.S. Coast Survey, and eked out a rather meager existence for awhile in Albany.

He was rescued in 1858 by the offer of a professorship in astronomy at Hamilton College, Clinton, New York (near Utica). He accepted, and also took command of its new observatory. This was equipped with a 13-inch refractor mounted in a cylindrical dome. With it he discovered one asteroid, (72) Feronia, in 1861, and two more, (75) Euridike and (77) Frigga, in 1862. It was still not easy for him to make ends meet; at one point, he had to consult with a lawyer about collecting his "last year's salary." Only in 1867, when a railroad magnate from nearby Delphi Falls agreed to provide funds for the astronomer's modest annual salary did Peters finally escape from his hardscrabble existence. The observatory was promptly renamed the "Litchfield Observatory," in honor of his benefactor (Figure 18).

At the time Peters was America's leading discoverer of asteroids. However, he would soon have a serious competitor in Watson, a much younger man whose salad days were yet to come. Indeed Watson was a remarkably efficient user of his time; he devoted all his off hours and summer vacations to ferreting out new asteroids among the stars. In 1868, he published his great book *Theoretical Astronomy Relating to the Motions of the Heavenly Bodies,* which established his reputation and won the praise of Le Verrier himself. That same year Watson discovered six asteroids—an unprecedented feat. For his achievements he was awarded the Lalande prize of the Paris Académie des Sciences in 1870.

Watson went on eclipse expeditions to Mount Pleasant, Iowa, in 1869 and then to Carlantini, Sicily, in 1870. He also traveled to Peking, China, for the transit of Venus in 1874. On October 10, 1874, he made the first asteroid discovery in China, which according to a contemporary account was "effected entirely through Watson's extraordinary recollection of the positions of the small stars in the neighborhood in which the planet was described."[11] As a courtesy to the Chinese royal family, he invited them to propose a name. They did so, calling it *Juewa,* the hope of China. On his return, Watson spent several weeks in Egypt and served as voluntary adviser to a geodetic survey of the country then being planned, for which he was named Knight Commander of the Imperial Order of the Medjidich of Turkey and Egypt.

Peters, too, was making asteroid discoveries and traveling to eclipses and transits. In Des Moines, Iowa, he observed the eclipse

18. **Litchfield Observatory, Hamilton College, Clinton, New York, c. 1875.** (Courtesy of the Mary Lea Shane Archives of the Lick Observatory.)

of 1869, an event he made memorable by his reference to Le Verrier's mythical birds.[12] For the 1874 transit of Venus he went to Queenstown, South Island, New Zealand. Without pausing for rest he renewed his quest for asteroids on his return to Clinton, discovering two on the night of June 3, 1875; he named them Vibilia and Adeona, after the minor Roman goddesses of journeyings and homecomings. He now had 22 to his credit; at the time Watson had 16.

The rivalry continued. Each man had his secret star charts of the ecliptic; each was intensely competitive and eager for fame. Their rivalry became personal, even bitter, and in due course erupted into a controversy seldom matched for venom in all the annals of the science.

fourteen

The Planet Hunters Come to Indian Country

The rivalry came to a head over the question of intramercurial planets. Basically, Watson was a believer, Peters a skeptic; Watson a supporter of Le Verrier, Peters otherwise. Interest had never been so intense as on the eve of the Great Solar Eclipse of July 29, 1878, the last total eclipse of the sun visible from the United States during the 19th century. Fanned by Le Verrier's recent transit predictions, public interest in Vulcan was at its height, and the eclipse seemed a good opportunity to settle the vexed question once and for all. It would be, in a sense, Vulcan's last stand.

Totality would last about 3 minutes and would start in northeastern Asia. The moon's shadow, 116 miles wide, would sweep across the Bering Strait into Alaska, strike Canada near its Pacific coastline, and then arc across the vast western United States. It would continue through Yellowstone and the majestic Wind River Range, on to the central highland of Wyoming Territory, down through Medicine Bow and south into Colorado, where it would cross the top of Long's Peak, fronting the foothills of the Rockies; through Boulder, Central City, Georgetown, Denver, Colorado Springs, and the snowy summit of Pikes Peak, then on to La Junta and Indian Territory (Oklahoma), and Texas, passing finally into the Gulf of Mexico between Galveston and New Orleans.[1]

194

Thanks to an $8000 appropriation from Congress, the U.S. Naval Observatory determined to establish eight observing stations along the shadow path from Montana Territory to Texas. Edward Singleton Holden, Simon Newcomb's protégé at the U.S. Naval Observatory, hoped to set up the northernmost outpost at Virginia City, Montana Territory, and invited Peters to observe the eclipse with him. "It is a great temptation," Peters replied, ". . . but I ought not to go, unless the Trustees . . . give me an assistant at the observatory—for which however there is probably little hope. . . . So, you go to Montana. Take care of not being scalped by the Indians!"[2]

The concern was real enough. The year before, the U.S. Army had been in action against the Nez Perces in that general region. Twelve months earlier, the Sioux under Crazy Horse and Sitting Bull had annihilated General George Armstrong Custer and elements of the U.S. 7th Cavalry at Little Big Horn. Now to the south, trouble was brewing among the Utes of northern Colorado. Against that background Holden acknowledged, "it was deemed best . . . that another station should be selected."[3]

Instead of Virginia City, the northernmost eclipse station would be at lonely Creston, Wyoming Territory, on the Union Pacific Railroad, which happened to follow a narrow strip of the eclipse path in the south-central part of the Territory. Newcomb, now Superintendent of the Nautical Almanac Office, proposed to go to Creston, as did William Harkness, another U.S. Naval Observatory astronomer, with a party made up of assistant U.S. Naval Observatory astronomer A. N. Skinner, famed telescope-maker Alvan Graham Clark, and French astronomer–artist Leopold Trouvelot.

On arrival, Skinner found Creston a "dull" place. Harkness described it as

a little bit of a hamlet, situated almost on the backbone of the continent, being only 2½ miles east of the divide which separates the watershed of the Atlantic from that of the Pacific. The country in its vicinity is flat and uninteresting. The ground is sparsely covered with sage-brush and coarse grass, trees are conspicuous by their entire absence, of brooks and streams there are none, and the horizon is bounded by the distant peaks of the Rocky Mountains. . . . Besides the station, water-tank, and coal-shed of the railroad company, the place contains only two small cottages, and its population consists of seven white adults, three children, and six Chinese laborers who keep the track in order.[4]

The congressional appropriation was generous enough to support many other parties. Dr. Henry Draper, physician and amateur astronomer of New York City, went to Rawlins, Wyoming, a larger stop along the same railroad line, which then boasted a population of 800 people and a good hotel. He went with Thomas Alva Edison of Menlo Park, New Jersey, the celebrated 31-year-old inventor of the phonograph. A trip out west would be a vacation for the hardworking Edison, but being Edison, he could not resist taking along one of his inventions—a *tasimeter*, he called it, actually a pocket-sized device for measuring infrared radiation. He hoped to use it to measure the temperature of the solar corona during the eclipse.[5]

Allegheny Observatory director Samuel Pierpont Langley climbed 14,147-foot Pikes Peak for the eclipse. Holden, with Johns Hopkins student and future Lick Observatory astrophysicist James E. Keeler as his assistant, ended up on the flat roof of the Teller Hotel at Central City, Colorado. Princeton solar astronomer Charles A. Young and English astronomer Arthur Cowper Ranyard established a station at Cherry Creek, near Denver. U.S. Naval Observatory astronomer G. W. Hill and New York amateur Lewis Swift, well known for his comet discoveries, observed from Denver itself. Asaph Hall, celebrated for his discovery a year earlier of the two small moons of Mars, went to the plains of La Junta, in southeastern Colorado. David Peck Todd established an observing station at Dallas, Texas, one of the southernmost sites along the eclipse's track across the United States.

Watson planned to observe the eclipse from Rawlins. A veteran of two previous total eclipses, he was intent on only one object: intramercurial planets. The quest absorbed him totally. He was 40, at the height of his powers, possessed of "a tireless energy ... that impressed every beholder," as a colleague, J. C. Freeman, later remarked.[6] Watson was portly; his maximum adult weight was 240 pounds, which led Ann Arbor undergraduates to dub him

"Tubby." To all appearances, by the standards of the time, he was a hale fellow, a man in vigorous good health and in the prime of life.

Even before Vulcan began to obsess him, he had begun a preliminary investigation into the possible existence of a planet beyond Neptune. He did so independently of the man who is usually mentioned in this connection, David Peck Todd, then an assistant under Newcomb at the Nautical Almanac Office. Todd's approach was to plot the residuals of Uranus's motion remaining in Le Verrier's 1873 theory, using a graphical analysis similar to that devised by Sir John Herschel after Neptune's discovery, to find the times of its conjunctions with a possible disturber. By this means he obtained approximate elements for a possible planet by August 1877 and mounted a brief, unsuccessful search for it with the U.S. Naval Observatory's 26-inch refractor on 30 nights between November 3, 1877, and March 6, 1878.[7] Though Todd has often been mentioned as a trail-blazer of transneptunian space, Watson's role has been completely forgotten. Our only knowledge of it comes from J. Norman Lockyer, who in writing from Rawlins in 1878 noted Watson "broke off work on a planet beyond Neptune to come to discover one inside Mercury."[8] Perhaps this is not too surprising; the idea was already in the air. A false rumor of a possible transneptunian planet had circulated as early as 1850–1851. Watson's admiration for Le Verrier may well have brought to his notice the latter's prediction made just 1 week after the discovery of Neptune, that one day a ninth planet would be found. In 1878 Watson's interest was overshadowed, permanently as it happens, by the spectacle and promise of the "Great Eclipse," as it came to be known.

His observing plan was simple. He would confine his observations during totality "to a search for any planetary bodies which might be visible in the neighborhood of the Sun." Indeed, Watson was "convinced of the correctness of Le Verrier's theory," and "believed that a diligent search, under favorable circumstances, would reveal the actual existence of one or more masses revolving nearer to the Sun than Mercury."[9] For this purpose, he secured a 4-inch Clark refractor from the Michigan State Normal School at Ypsilanti (now Eastern Michigan University).

After finishing up some last-minute business, Watson ordered a piano from William Steinway, New York, then set out on July 22. That night he and Mrs. Watson stood on the platform of Ann Arbor railroad depot about to begin their journey to the West and the tribal lands of the Ute and Wind River Shoshone, a journey of unexpected consequences that would forever change the course of their lives. They were bound for Rawlins, Wyoming Territory, the turbulent frontier town with a population of 800, created 9 years earlier by the Union Pacific Railroad.

On arrival, the first thing to strike Watson was the remarkable transparency of the sky. He found it easy to make out the satellites of Jupiter with the naked eye—or so the (not unknown to exaggerate) local papers reported. Rawlins was bustling with astronomers (Figure 19). Watson met up with J. Norman Lockyer, a noted British solar physicist and founder and editor of the science journal *Nature.* Lockyer planned to join Simon Newcomb at Separation, some 13 miles further west, where totality was expected to last for 2 minutes 56 seconds. He had hired a railcar specially fitted out as a photographic darkroom in which he would be able to develop his wet-process plates of the coronal spectrum.

As at the 1869 eclipse, Newcomb (Figure 20) had intramercurial planets before his eye, and planned to search for them with a 5-inch equatorial. As mentioned earlier, his original plan had been to occupy a site at Creston. However, he had taken the precaution of authorizing his advance guard, Commander W. T. Sampson and Lieutenant C. G. Bowman, United States Navy, to select another in case he found the former "impracticable from any cause whatever."

Sampson was a man of considerable derring-do. He had graduated at the head of his class from Annapolis in 1861 and survived a torpedoing during the American Civil War. Later he would take command of the United States fleet at Santiago Bay in the Spanish–American War. His preparations for the eclipse began July 9 with the authorized purchase of supplies from Ft. Fred Steele, a military post established in June 1868 as part of the protective system for the Union Pacific and the Overland Trail. The post was located about 15 miles east of Rawlins, on the east side of the North Platte River close to the railroad bridge spanning the river.

19. Thomas A. Edison with party of astronomers at Rawlins, Wyoming, before observing the eclipse of the Sun July 29, 1878. Left to right: Professor George F. Barker, Robert M. Calbraith, Dr. Henry Morton, not identified, —Meyers, D. H. Talbor, M. F. Rae, Marshall Fox, Professor James C. Watson, Mrs. A. H. Watson, Mrs. Henry Draper, Dr. Henry Draper, Thomas A. Edison, J. N. Lockyer. (Courtesy United States Department of the Interior, National Park Service, Edison National Historic Site.)

20. Simon Newcomb (1835–1909). (Courtesy Mary Lea Shane Archives of the Lick Observatory.)

Major Thomas Tipton Thornburgh, officer commanding, gave every possible assistance and dispatched an officer to supervise the setting up of the eclipse camp. He also supplied "a detail of seven men, including a carpenter and a mason and a team," which with the requisite tents, went on ahead to Creston. Sampson and Bowman arrived there on July 15. "It was then blowing half a gale of wind," Sampson later reported, "so that it seemed that it would be impossible to use our instruments in case of so much wind on the day of the eclipse. Upon inquiry I learned that the strong westerly winds were nearly continuous."[10] When the winds failed to subside, Sampson and Bowman decided to check conditions at Separation, an isolated Union Pacific rail stop at an elevation of 6,901 feet, located about midway between Creston and Rawlins, some 15 miles from the summit of the Rocky Mountains. The town derived its name "from the fact that at this place the various parties of surveyors who had been together or near each other for the last hundred miles, separated to run different lines to the westwards."[11]

Soon after 1878 Separation died, abandoned when Union Pacific shifted its track south. Other than a few metal spikes and other items of inhabitation, no trace of it survives. John Jackson Clark, who was station agent at the time, reminisced in 1926 that it was in the valley, about 13 miles west of Rawlins. It consisted only of a wooden water tower, a rail siding, and a few small wooden buildings where the station agent lived and worked. It was a place bleak and forlorn, surrounded by a rough alkali plain, broken by nothing save the avenue of telegraph poles that marched alongside the railroad track.[12] But at least there was accommodation for the astronomers.

About ¾ of a mile east of the station, Sampson and Bowman discovered a semicircular sand dune, "15 feet high, which furnished an admirable protection to the southward and westward. This bank was quite abrupt on the east side and covered with low bushes." Thus it furnished at least some shelter from the wind. Here, then, at latitude 41° 45' 51".14, N and longitude 107° 27' 02".55, W of Greenwich, they decided to establish their tent camp. The soldiers began the construction of brick piers on which the instruments could be set up.[13] Work got underway on Tuesday, July 16, the day Newcomb started out from Washington. He was informed of the change of plans, and arrived at Separation 5 days later. Edison, too, came out briefly to stare at the "government astronomers." Finding everything in order at Separation, Newcomb returned to Rawlins to await the momentous event. It was there he met up with Lockyer, and invited him to join his party for the eclipse.

For 2 weeks prior to the critical date the weather was thoroughly unpromising, with clear mornings and cloudy afternoons. Hope rose when it was noticed that each day the cloud tended to form at a later time. Torrential rain fell for 3 hours the day prior to the eclipse, and masses of dark clouds raced across the plains. The next day, Monday July 29, the sky throughout Wyoming was perfectly clear—"as slick and clean as a Cheyenne free-lunch table."[14] Trouvelot, with his artistic eye, wrote from Creston that the "morning . . . was a glorious one; the sun rose clear and bright above the distant horizon of the great alkali plain; not a cloud was to be seen in the deep-blue sky."[15]

Nevertheless the wind got up early in the morning, and soon blew with galelike force. At 8 o'clock, Trouvelot recorded that while taking breakfast under the tent "we found ourselves and the dishes completely covered with sand and dust." He had dug an underground observatory, and at 9 o'clock, when he commenced observing the Sun, "thick clouds of hot sand and dust" were falling in on him from every direction. "The most violent gale we had yet experienced began to blow from the west, and increased in intensity until nearly the time of the eclipse," Newcomb noted at Separation. "Its most deleterious effect was filling the air with an impalpable dust, and thus diminishing the remarkable clearness of the sky. This dust seemed to act not so much by cutting off the sunlight as by forming a halo around the Sun."[16] Watson found "everything . . . favorable except that we were annoyed by brisk wind at times, which blew the dust down into our sheltered recess."[17]

As the eclipse approached, Newcomb thought it worthwhile to further protect the instruments. He asked the military to place several sections of snow-fence on top of the bank as a further wind break. Such assistance was crucial. He later acknowledged thanks to William Tecumseh Sherman, General of the Army. Without the help of Major Thornburgh and his command, Newcomb said, "occupation of the station would have been impracticable except at great expense." A stark reminder of the risks run by the astronomers can be gleaned from Newcomb's report, dated October 16, 1879. Little more than 2 weeks before, unrest among the White River Agency Utes of northwest Colorado erupted in violence. Ordered south to quell the uprising, Thornburgh was killed at the outset of a sharp and bloody engagement at Red Canyon, near Milk River on September 29, south of Separation.

Watson originally planned to observe with Lockyer. On the morning of eclipse day, he and Mrs. Watson left their accommodation in Rawlins and traveled to Separation aboard the photographic railcar Lockyer had hired from one J. B. Silvis. However, after arriving at Separation, as Lockyer later recalled, "we found Prof. Newcomb, whose camp was about a mile away, and it was then agreed that as both he and Prof. Watson were to hunt for the planet they had better be together, so I lost his company during the eclipse."[18]

Although mounted equatorially, the 4-inch Clark was not furnished with setting circles. However, Watson explained:

> This . . . was not of consequence in the present instance, because I had determined not to undertake to read the circles for any strange object which I might find, but to adopt a plan which offered greater facility and might give positions with all necessary precision. My plan was to place upon the axes, circles covered with white card board, and to place near the face of each circle a pointer with a knife edge vertical to its plane. The position of any object observed could then be readily marked with a pencil, and the angular measures could be subsequently read by mounting these circles on the axis of a graduated circle. To read divided circles would require considerable time, while the pointings can be marked on the paper disks in a few moments. And besides, while a doubt might be raised as to the correctness of recorded circle readings, no such doubt can exist in reference to the positions marked on these circles.[19]

According to Donald E. Osterbrock, "the tension of eclipse observations in which delicate scientific equipment must work at a particular moment of time completely outside the observer's control . . . is unimaginable."[20] More than one astronomer has had nightmares on eclipse expeditions. All that one can do is make one's preparations as well as possible, then await, with as stoic an attitude as possible, one's fate. At Separation, the astronomers put the last touches to their instruments. By 2 p.m., when the Moon took the first small bite out of the Sun, everything seemed in readiness. As the partial phases proceeded, Newcomb withdrew to a dark room, in order to dark adapt his sight. He did not emerge until 3 minutes before totality, when he found the "lurid color of the landscape . . . very striking. The light seemed no longer to be that of the sun, but rather to partake of the character of an artificial illumination."[21]

The last sliver of sunlight disappeared behind the Moon in the shimmer of Baily's beads, a phenomenon caused by irregularities of the moon's limb that break up the disappearing shred of sunlight into specks. Totality came at just after 3:16 p.m., not without the usual confusion. The gusting winds began to fall. The dusty halo that had earlier bothered Newcomb vanished. In the eerie stillness, the pearly white corona came out with startling abruptness (Figure 21). Newcomb spent a minute transfixed by it. At last, he summoned the self-discipline and sense of purpose needed to conduct a systematic search for intramercurial planets. A few bright stars were visible at a glance: Procyon, Vega, Arcturus. Venus was prominent low in the western sky, where a few clouds took on a peculiar ruddy luster. Mercury, Mars and Regulus were all bunched together near the meridian.

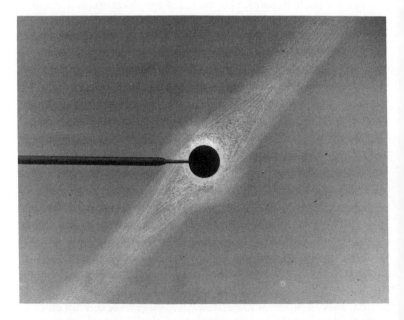

21. The Great Solar Eclipse July 29, 1878, as seen at Separation, Wyoming, by Simon Newcomb, U.S. Naval Observatory party.

Newcomb began sweeping with his small telescope east of the Sun. His first impression was "the sky was so bright that it would be very easy to overlook a faint object unless the eye looked directly at it."[22] Before him was a copy of a map published with the U.S. Naval Observatory Eclipse Instructions. It contained all stars near the Sun down to the 7th magnitude. His first sweep covered a swath of sky between 18° and 19° declination, out to several degrees from the Sun. He recognized only the stars on the map. He then began sweeping north and south of this, but with the same negative result. Now the end of the eclipse was fast approaching, and he began making even wider sweeps, "at random, with the object of picking up some object by chance." As he did so, he captured a star, and tried to measure its position using his setting circles.

Watson, meanwhile, had been proceeding with his own plan. From previous eclipse work, he had decided it was best not to try to sweep too large an area. He confined his attentions to a search zone only 15° long and 1½° wide, having previously committed to memory the relative places of all stars in this zone down to seventh magnitude—a mnemonic *tour de force*, but no greater than the usual feats he performed in his calculations. As an added check, he placed the U.S. Naval Observatory reference chart before him.

He began sweeping even before totality, using a 45X ocular. He found no stars within 8° to 15° east or west of the Sun. At the onset of totality, he placed the Sun in the middle of the field and began moving the telescope, slowly and uniformly, toward the east. He retraced his path, then moved the telescope one field to the south and began sweeping again. He encountered δ Cancri and other known stars but nothing unusual. Once again he centered on the Sun, and began sweeping to the west. Between the Sun and θ Cancri, he came across "a ruddy star whose magnitude I estimated to be 4½. It was fully a magnitude brighter than θ Cancri, which I saw at the same time, and it did not exhibit any elongation, such as might be expected if it were a comet." He attempted to obtain its position. "My plan did not provide for any comparison . . . with a neighboring star by micrometric measurement," he later recalled, "and hence I only noticed the relation of the star to the sun and θ Cancri. Its position I proceeded at once to record on my circles . . . and I recorded also the chronometer time of observation. This star was denoted by *a*."[23]

After marking the position of this object and the Sun on his paper circles and assuring himself that the pointing of the telescope had not been disturbed in any way, he extended his sweep for several degrees further west of the Sun. There he encountered yet another, even brighter star, "also ruddy in appearance, which arrested my attention." He marked the position of this object, *b*, on his circle, and ran over to Newcomb "in hopes that he might, before the sunlight became too bright, get a place of the strange star which I had first observed. . . . But he was occupied in reading his circles for a star which he had in the field, and consequently his instrument could not be disturbed. He said that the star whose position he was seeking was north of the Sun. I stated to him that I had found two strange objects, both of which were south of the Sun; that one might be a known star on the chart, but that the other was certainly a stranger."[24] The star that Newcomb was busy measuring with such diligence proved to be an ordinary star, and he could not help reflecting afterwards: "It is of course now a matter of great regret that I did not let my own object go and point on Professor Watson's."[25] (Figure 22.)

By the time Watson had returned to his telescope, the Sun had peeked out from behind the Moon's limb, and the 3 minutes of totality were over. The end no doubt seemed abruptly premature. There is an almost universal sense, at all eclipses, that totality passes in a small fraction of its actual duration.

Against the suddenly brightened sky, Watson was unable to recover his second object. Therefore he was unable to say for sure whether it was a known star (ζ Cancri) or not. About the first ruddy object, near θ Cancri, he was sure; it was not a known star, and he had the precious record of his paper circles, carefully inspected by Newcomb and Lockyer in situ, to prove it.

Other planet-hunters now sprang into action as the Moon's shadow raced toward them. Holden searched at Central City with his small telescope, but telegraphed: "Whole eclipse perfectly observed. I find no Vulcan as large as sixth magnitude." Professor Elias Colbert of Chicago also failed to see Vulcan from Denver. As-

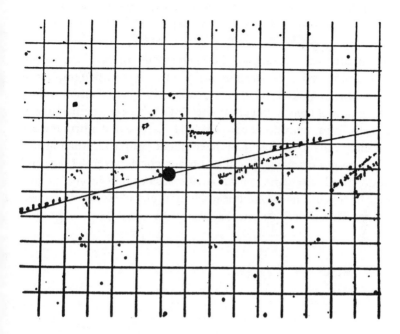

22. Watson's chart of the 1878 Eclipse showing his supposed discovery of Vulcan at the time of the eclipse, a little to the right of the Sun. (Reprinted with permission from Michigan Alumnus, July, 1938.)

aph Hall, assisted in his Vulcan search by O. B. Wheeler, sent another negative report from La Junta. His report in particular gives a good feel for the circumstances and technique:

> The day at La Junta was all that could be desired for observing a total eclipse. My own special work during totality was searching for an intramercurial planet, the supposed Vulcan, indicated by the researches of Le Verrier on the orbit of Mercury. Before the eclipse I studied the configuration of the stars as they are laid down on the chart published by the Observatory [U.S. Naval observatory, Washington D.C.], and during totality a copy of this chart was placed a few feet in front of me, so that I could refer to it instantly. As soon as totality began I turned my shade to the free opening, and commenced sweeping above the Sun and near the ecliptic. My sweeps extended from the brighter part of the corona to a distance of about

ten degrees from the Sun. The magnifying power was so great that the sweeping could not be done very rapidly. In this part of the sky I saw nothing but the stars laid down on the chart.[26]

Before the eclipse, plans had been made, in the event Vulcan was discovered at Separation, for a Union Pacific telegrapher to send a message to the Dallas railroad station. There a man waited on horseback, ready to gallop with Vulcan's position the quarter mile to Todd, who stood ready at his telescope to confirm the observation. In the confusion of the moment, people lost their heads. No message was sent. The man on horseback never set out, and Todd could only carry out his own search as best he could. That day he telegraphed: "No intramercurial planet seen with comet-seeker. Thin clouds. No stars seen near the Sun."[27]

fifteen

Dr. Peters Confronts the Wild Geese

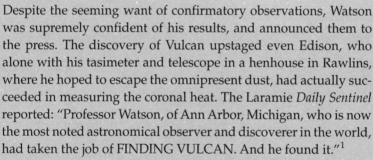

Despite the seeming want of confirmatory observations, Watson was supremely confident of his results, and announced them to the press. The discovery of Vulcan upstaged even Edison, who alone with his tasimeter and telescope in a henhouse in Rawlins, where he hoped to escape the omnipresent dust, had actually succeeded in measuring the coronal heat. The Laramie *Daily Sentinel* reported: "Professor Watson, of Ann Arbor, Michigan, who is now the most noted astronomical observer and discoverer in the world, had taken the job of FINDING VULCAN. And he found it."[1]

Lockyer, on August 1, telegraphed information about Watson's discovery to Airy at Greenwich and Admiral Ernest Mouchez, Le Verrier's successor at Paris. His telegram included the object's magnitude, 4½, and a tentative position: R.A. 8h 26m, declination 18° 0′ north. This put it some 2½° southwest of the Sun. Airy, replying August 3, noted it was within a degree of θ Cancri (the latter's position he gave as R.A. 8h 24m 40s, declination +18° 30′ 20″). He added, "it may be worth remarking that . . . there may be a possibility that the object observed is in reality this star. . . . The magnitude of this star is, however, smaller than that given by Mr. Watson. . . . This discrepancy may very easily occur in the hurry of such a sensational observation, as on these occasions the time at the disposal of the observer is so limited."[2] Mouchez hinted the same possibility.[3]

The reaction soon became more widespread. The London *Times* announced the evidence for the existence of Vulcan could not have come from a more reliable source. Although Watson's planet was not large enough to explain the advance of the perihelion of Mercury, there was no doubt other small planets remained to be discovered; "future eclipses alone can resolve the question." A brief essay in Lockyer's journal *Nature* reflected that as θ Cancri was the only star near the position indicated by Watson, "it remains to be seen from further intelligence whether there was any possibility of the star having been the object noted; if it were separately remarked, or if the observed position does not admit of such change as would be necessary for identification, then it may truly be said that the American astronomer will have rendered the occasion of this eclipse a memorable one in the history of science."[4] Daniel Kirkwood, of the University of Indiana, who had predicted additional planets of the solar system would turn up ever since Neptune had been discovered, wrote in *The Popular Science Monthly:* "It is not improbable that the detection of Vulcan may be merely the first in a series of similar discoveries. . . . The prospect of planetary discoveries in this part of the system is at present more hopeful than in the space beyond the orbit of Neptune."[5]

Even before more satisfactory information was forthcoming from Watson himself, there was a startling new development. Lewis Swift, a skillful amateur astronomer from Rochester, New York, acclaimed for his discovery of several comets (Figure 23), produced independent observations of the intramercurial planet from Denver, where totality had begun some 5 minutes after it had done so at Separation. Swift had accompanied Colbert's party, but carried out his own search. The reason his discovery was not immediately reported by the Associated Press was, according to a story in the *Rochester Democrat*, "chiefly owing to Prof. Swift's modesty in heralding the results of his labors and his desire to carefully determine the significance of his observations before making public announcment." According to more skeptical opin-

23. Lewis Swift (1820–1913). (Courtesy of the Mary Lea Shane Archives of the Lick Observatory.)

ion, however, Swift had waited for Watson's report, which came the day after the eclipse, in order to "harmonize it" with his own.

Like Watson, Swift had carefully planned his search over several months ahead of the eclipse. Despite the planet having been seen only by the eye of faith, he had resolved "to devote at least two minutes of totality to a rigid search for the hypothetical Vulcan, in whose existence I, for many years, have cherished a hope."[6]

Swift's instrument was the same 4½-inch refractor he had used to discover new comets in 1877 and 1878 (the latter on July 8, just 3 weeks before the eclipse). He set the instrument on a low post, and instead of using an observing chair, spread a carpet for himself on the tufts of buffalo grass. This excited "much comment by my companions." He fully intended to observe the eclipse by lying on the ground, since from his comet-sweeping experience he had found this to be the most comfortable. In his view, it gave him "a great advantage over the other members of the party, who were obliged to assume constrained positions, which tended to un-steadiness of vision."[7]

As the eclipse approached, his telescope was shaken by fitful gusts of wind from the southeast. To overcome this, he applied an-other trick learned from comet-hunting: he tied one end of a ten-foot pole to the telescope near the eye-piece, and braced the other end against the ground. He fully planned to untie the telescope from the pole as soon as totality began, but in the excitement and preoccupation of the moment, forgot to do so; "a blunder to which," he declared afterwards, "I owe the discovery of a stranger."[8] Because of the pole, he was unable to move the tele-scope east of the Sun without plunging the pole deep into the buf-falo grass. Thus his field of view was confined to a small area west of the Sun. As he proceeded to sweep this region, he "ran across two stars presenting a very singular appearance, each having a round red disk and being free from twinkling." Their position was, he estimated, about 3° southwest of the Sun. "I saw them twice," he noted, "and attempted a third observation, but a small cloud obscured the locality." Before the small cloud vacated this position, totality was over. Swift concluded in his report: "The stars were both of the fifth magnitude, and but one is on the chart of the heavens. This star I recognised as Theta [θ] in Cancer. The two stars were about eight minutes apart. There is no such config-uration of stars in the constellation of Cancer. I have no doubt that the unknown star is an intra-Mercurial planet. . . . The star I saw may have been the same that was seen by Prof. Watson."[9] He men-tioned, in passing, Colbert's unsuccessful search for the planet; but noted the latter's "field had not been very large."

Swift's testimony, added to Watson's, seemed to establish the existence of Vulcan beyond all possibility of doubt. Watson wrote to Lockyer from Ann Arbor on August 14: "You will be pleased to hear that the planet was seen a few minutes afterwards by Mr. Lewis Swift. . . . I do not know whether he obtained anything more than an estimate of the position; but the place in which it is reported that he saw the planet agrees with my observation. This corroboration is peculiarly fortunate, considering the negative results of other observers."[10]

So far the basic facts seemed definite enough, though there were a few refinements in the details. After placing the paper circles used for marking his pointings for the planet and the Sun on a properly graduated circle, Watson published a revised position for his first object: R.A. 8h 26m 54s and declination +18° 16'. At the same time he clarified in a letter to Airy his object had been "very much brighter than θ Cancri, which was seen a little further to the west." Even with a magnification of only 45×, he emphasized, it had had a "perceptible disk." Thus there had been nothing starlike about it, and "there was no elongation such as might be expected if it were a comet."

After this, Airy was convinced that Watson had not merely recorded θ Cancri, but still wondered whether he had possibly seen a comet. "Prof. Watson's statement appears to render it very highly probable that the object seen is really an intra-Mercurial planet," wrote the Astronomer Royal. "I remark, however, that the reason for excluding the supposition of its possible cometary character does not seem quite conclusive, as, when the tail of a comet and the small appendages of its head are invisible, the nucleus is usually circular."[11]

For that matter, despite Watson's object having been widely hailed as Vulcan, the identification with the planet of Le Verrier and Lescarbault had been assumed rather than established. Now Gaillot, Le Verrier's former assistant, informed Watson that he had carried out extensive calculations, from which he concluded the position of the object seen by Watson was indeed "in accordance with Dr. Lescarbault's discovery, so long denied by M. Le Verrier's opponents."[12] Mouchez noted with satisfaction the discovery had consecrated the splendid scientific endeavors of Le Verrier.[13]

These were exhilarating times. Once more Vulcan seemed to be carrying all before it. "The planet Vulcan, after so long eluding the hunters . . . appears at last to have been fairly run down and captured," trumpeted *The New York Times*. "At least it seems to us that the observations . . . must for the present be taken as conclusive, though perhaps not settling the question beyond the possibility of re-opening or dispute."[14]

By late August a cloud had formed on the horizon. Watson discovered an error in the correction applied to Newcomb's chronometer. This led to yet another revision of the position of the planet. For July 29, 1878, 5h 16m 37s Washington Mean Time, it became: R. A. 8h 27m 35s and declination +18° 16'. The probable error in this position was estimated at only 5 arc minutes.

This was a minor detail. But Watson at the same time made another, and quite surprising, announcement. "The more I consider the case the more improbable it seems to me that the second star which I observed and thought might be ζ Cancri, was that known star. I was not certain in this case whether the wind had disturbed the telescope or not. As it had not done so in the case of any other of six pointings which I recorded, it seems almost certain that the second was also a new star."[15] In other words, there seemed to be two planets, not just one.

Meanwhile, C. A. Young, Princeton's distinguished solar astronomer, was the first to express reservations about Watson's observations in a note in the *American Journal of Science and Arts*. Although Young had been intent on observing the solar corona at the eclipse, he noted that Newcomb, Holden, Hall, and others had all gone over apparently the same ground as Watson and had found nothing. Watson protested Young's "inference that they obtained negative evidence of some value to dispute the discovery announced."[16] Instead, he wrote, careful attention to the details of those other searches proved otherwise. Newcomb had scanned the region north of the Sun, hence he could not have observed either of Watson's objects. Holden, who had never witnessed a total eclipse before, confessed he had greatly underestimated the illumination of the sky. Thus he had observed "with optical power insufficient for a search under these circumstances." With his hand-held comet-seeker of 2½ inches aperture, he had twice swept over a space of 30° long and 10° wide, finding no stars until he reached as far out as Regulus and Mars; his negative result was of little value. Hall's opinion was not to be lightly set aside. Like Newcomb, he had confined his search north of the Sun, deferring to his assistant Wheeler for the south. Both had used an inconveniently high magnifying power of 160×, so "the sweeping could not be done very rapidly."[17] Watson noted "anyone who has tried to find so bright an object as Jupiter by attempting to direct the telescope, with a high power, without using the finder, will know how uncertain a search would be under the circumstances."[18] Colbert had also used too high a magnifying power, while Todd's search was largely vitiated by cirrocumulus clouds passing in front of the Sun, so dense as to obscure the corona considerably and to blot out stars as bright as δ Cancri.

The negative observations were, in Watson's view, entirely inconclusive. Watson continued to regard Swift's observations as furnishing "important corroborative evidence," but he did not base his case on them. "The records of my circles cannot be impeached by all the negative evidence in the world," he wrote decisively. "There are no known stars in the places which they give, and hence I cannot be mistaken as to the identity of the objects which I observed."[19]

Swift was always Watson's understudy in the whole affair. In early September, he, too, began to furnish additional information about his observations. He wrote to *Nature:*

It may be well here to state that I was prevented from searching to the east of the Sun, in consequence of forgetting to untie a string with which I had tied, to the eye end of the telescope, a long pole to prevent the wind from shaking it, the end resting on the ground not allowing the instrument to be moved to the eastward. It is undoubtedly to this circumstance, which at the time seemed untoward, that I owe the discovery of Vulcan. . . . Almost at once my eye caught two red stars about 3° south-west of the Sun, with large, round, and equally bright discs, which I estimated as of the fifth magnitude. . . . I then carefully noted their distances from the Sun and from each other, and the direction in which they pointed, &c., and recorded them in my memory, where, to my mind's eye, they are still distinctly visible. I then swept southward, not daring to venture far to the west for fear I should be unable to get back again, and soon came upon two stars resembling in every particular the former two I had found, and, sighting along the outside of the tube, was surprised to find I was viewing the same objects. Again I observed them with the utmost care, and then recommenced my sweeps in another direction, but I soon had them again, and for the third time in the field. This was also the last, as a small cloud hindered a final leave-taking just before the end of totality. . . . I saw no other stars besides these two, not even δ, so close to the eastern limb of the Sun.[20]

Swift's objects were 7 or 8 arc minutes apart, and located so that a line drawn between them would point to the center of the Sun. He assumed that one of them was θ Cancri, the other the new planet.

"Thus far," he concluded, "all seems clear and free from doubt, but it is just here where the trouble begins, for, unfortunately, I could not tell which was the star and which the planet. Happily Prof. Watson comes to the rescue, and with his means of measuring, says 'the planet was nearest the Sun.' "[21] Equipped with this supplemental information, Swift succeeded in working out the position of his "planet," and found that it agreed to "a close approximation" with that given by Watson. At the time, he had not yet received Watson's final position. The agreement with the latter was not so satisfactory. Also, in reducing his observations, Swift had made a serious mistake, which, when corrected, greatly increased the difference in right ascension between his and Watson's objects. Swift confided to Watson his trouble in reconciling the observations, but the trouble was not with the positions alone, since Swift's description of two equally bright stars stood in open contradiction to Watson's statement that his object was much brighter than θ Cancri.

The situation was now becoming awkward. There seemed to be no choice but to admit Swift and Watson had seen different objects after all (Figure 24). Moreover, as Swift later summed up, it was necessary to accept "the two objects seen by me were both intra-Mercurial planets, and that I did not—as was for a time supposed—see θ Cancri. Prof. Watson saw θ, and, some 42 minutes of arc southeast of it, another planet, and determined its position, and near to ζ Cancri still another, whose position also he fortunately determined, making four in all."[22] Instead of one planet or even two, Watson and Swift were now claiming four planets, much to the astonishment of the astronomical world at large.

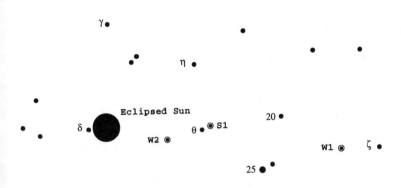

24. Diagram of supposed Watson/Swift objects July 29, 1878.
W1 and W2 represent the objects seen by Watson S1, that seen by Swift.
the Sun was in the constellation Cancer, a small distance south of the open
cluster Praesepe. West is to the right, east to the left, north at the top.

C. H. F. Peters had followed events with growing skepticism.
He did not share the universal approbation exemplified by another Peters, C. A. F., editor of the distinguished journal *Astronomische Nachrichten,* who after receiving some of Watson's earliest communications about the discovery,[23] congratulated him:
"Thank you, Sir, obligingly for the kind communications about
your discovery of Intra-Mercurial Planets, and I take the liberty to
congratulate you upon this discovery, which doubtless is of high
importance for the Science. It is [a] pity that Le Verrier has not
lived so long as to see this confirmation of his supposition about
the cause of some disturbance in the motion of Mercury."[24]

C. H. F. Peters never had any faith in the existence of intramercurial planets, and even now, when privately asked whether
he was ready to accept them, retorted, "No, I want to see them myself!"[25] He was sharpening his weapons, and in early 1879
mounted a devastating attack in the *Astronomische Nachrichten.*[26]
His paper, "Some critical remarks on so-called intramercurial
planet observations," was described by Joseph Ashbrook as "a
strange blend of sharp insight and utter tactlessness."[27]

Of all the observers who searched for Vulcan at the 1878 eclipse, only two, said Peters, reported seeing anything unusual, which "without much hesitation they pronounce to be intramercurial planets." Peters felt of the two, Swift's could be given comparatively short shrift, since "in his successive publications is perceivable so singular a gradation in statements." In the end, he decided only Watson's were really deserving of a serious scientific discussion.

Peters's main point was that the Ann Arbor astronomer had overestimated the accuracy of his paper circle method of measuring the positions of his stars. "All this is very well and ingenious," he admitted, but then filed quickly to a point. "The marking was done in the dim light of the total eclipse, or with lamplight. Either the slightest touch would bend the pointer, the flexible brass wire, a little to the side, or a parallax of some amount was unavoidable. The marking had to be done expeditiously and with a certain hurry." Under the circumstances, errors of considerable magnitude were inevitable. After examining Watson's diagram, Peters concluded this is what happened. On Watson's 5-inch diameter circle, he would have had to distinguish gradations of 1/275-inch in order to confine his limit to what he claimed: 5 arc minutes. But at best, said Peters, he could never have distinguished in the subdued light of totality gradations of less than 1/70 inch, which would have led to a probable error of 20 arc minutes. θ Cancri then falls within the circle of probable error. "It is . . . quite apparent to every unbiased mind, that Watson observed θ and ζ Cancri, nothing else." Watson had expressly stated in one of his communications to Peters that he had seen both his first planet *a* and θ Cancri. But according to Peters, this remained doubtful, since he had not seen them at the same time. In a reference to Watson's earlier statement ("the records of my circles cannot be impeached by all the negative evidence in the world") Peters protested: "To be sure, there are no known stars in the places indicated. But records of circles, and star places therefrom derived, are imbued with probable errors, and within the limits of these errors Prof. Watson can well be mistaken as to the identity of the objects."[28]

Peters advanced various possible explanations why θ Cancri, otherwise a white star, should have appeared ruddy to Watson. (Perhaps the sand dune under which Watson set up his telescope was responsible; perhaps it was an effect due to the corona.) As for its appearing as a small disk, Peters suspected this may have been due to wishful thinking on the observer's part, and that "the disks were somewhat imaginary." Swift's observations of small disks came in for even rougher treatment: "As Watson's careful sweeps did not show such a pair of stars, it is difficult to imagine where they were. May be, that Mr. Swift while endeavoring to release his telescope from the strange yoke tied to it, moved it off to a quite different star group, or perhaps also that the strain of the ten feet pole disturbed temporarily the glasses, and made him see double. Anyhow, Prof. Watson's and Mr. Swift's observations do not confirm . . . but rather contradict each other."[29] What is more, neither Watson's nor Swift's bodies were able to explain the theoretical advance of Mercury's perihelion found by Le Verrier, which indicated an unknown mass equal to that of Mercury. To the contrary, based on what Peters regarded as reasonable assumptions, "between one million and 38 millions of Watson's or Swift's bodies would be required to make up the mass of Mercury."[30]

Peters closed his paper with a wide-ranging attack on all the observations ever reported of dark spots in front of the Sun. Lummis's object was, he demonstrated from his own observations, but two sunspots confused with one another. Likewise, most of the other reports seemed to have been nothing more than short-lived sunspots. He admonished, in his rather broken English prose:

With few exceptions, only persons otherwise wholly unknown as astronomical workers have been favored with a view of the mythical planet. It never has revealed itself to any of the astronomers, that, now about during half a century, have kept a continual watch of the Sun, from the, in every respect very reliable Schwabe on, who, as he himself tells, began his long and memorable scrutiny of the Sun with the express purpose of looking out after a planet. Since the time of Carrington all the spots have been counted, measured, photographed, but never a trace of 'Vulcan' has been found by the astronomers engaged in that work. Much is said about negative evidence being insufficient; but then, in this case, tacitly is assumed that simple positive assertion is equal to positive evidence. Not one, however, in the whole series of cases affords a proof, that really an intra-Mercurial planet had been observed. Most of them even can not be called observations, in an astronomical sense, consisting only of rude general statements, without precision, often contradictory in themselves, always incomplete.[31]

The data of observation, the "pillars" of Le Verrier's calculations, have, Peters noted, under close scrutiny "slipped away entirely."[32]

Peters's insightful criticism found its mark. Vulcan did not exist; therefore Watson and Swift had been mistaken. "The most feasible explanation of the puzzle," summed up Agnes M. Clerke, "seems to be that Watson and Swift merely saw each the same two stars in Cancer: haste and excitement doing the rest."[33]

Indeed, Peters's version of what happened quickly gained credit in the astronomical world. "I thank you very much for your papers on the intra-Mercurial planet," wrote C. A. Young from Princeton, New Jersey, June 2, 1879. "On the whole it looks as if you were right, and as if Prof. Watson had overstated the reliability of his observation. As for Swift's, I have never taken much stock in it. I am disposed to think him honest but I imagine he was deceived by some reflection from the glasses of his instrument."[34] Astronomers had been burned too often by Vulcan to give it their unqualified belief, while the inconsistency of Swift and Watson's observations seemed difficult, if not impossible, to reconcile.

Still the controversy was far from over. Feeling savaged by Peters, on May 15, 1879, Watson responded in the same publication. He strongly protested Peters's "misstatement of the facts," and began with a defense of the accuracy of his pointers. In particular Peters's charge that they were so elastic as to give way several degrees under the touch of a pencil was "wholly untrue." He added, "it seems to me the grossest of unfairness to attempt to discredit an observation made by an experienced observer by deliberately misrepresenting the circumstances of the observations." Watson further stated:

> Professor Peters's whole attack upon the integrity of my observations is not of the slightest consequence, since he has created the errors in his own brain and has then pro[ceeded] to assail them. I do not intend to engage in any controversy about these matters and especially with a person who was, at the time of the observations,

more than 2,000 miles away from the place where the eclipse was observed. I repeat here the emphatic declarations:

1. I observed, during the total eclipse of July 29th 1878, a new star between θ Cancri and the Sun, and south of the Sun, whose position and magnitude were as already published by me.
2. I observed another star, which I believe to be a new star, whose magnitude and position were as already published by me.[35]

Watson concluded:

I do not propose to discuss the integrity of my observations pointing to the existence of intramercurial planets with either Dr. Peters or Mr. Flammarion, both known enemies, and assailants of the late, illustrious Le Verrier. . . . I have been engaged in making astronomical observations, observations of precision, for 23 years, and being fully cognizant of every circumstance connected with the observations in question, and having observed, with deliberation and with care . . . it matters not to me what these men think or what motives prompt their action, I know whereof I affirm. Whether or not the two new objects which I observed were intra-Mercurial planets I cannot positively assert; but I certainly have the right to express my honest belief that they are. I hope to be able to give, ere long, good reasons for the faith that is in me.[36]

As he wrote this, Watson was preparing to move from Ann Arbor to Madison, Wisconsin. He had been lured west by University of Wisconsin president John Bascom, to take charge of its new Washburn Observatory (Figure 25). The latter had been financed by Cadwallader C. Washburn, lumber and milling baron, six-term member of the U.S. House of Representatives, Civil War major general, and governor of Wisconsin from 1871 to 1873. Watson's only condition had been that the observatory be furnished with an instrument "equal or superior to that of the observatory of Harvard University, Cambridge."[37] It was. The 15.6-inch Clark refractor arrived in Madison January 1879. It had narrowly surpassed the 15-inch Merz refractor at Cambridge, and was the third largest instrument of its kind in the United States, behind the U.S. Naval Observatory's 26-inch refractor and Dearborn Observatory's 18-inch refractor. Erected on a lovely site near Lake Mendota, it was isolated from Madison by the university campus and set in an idyllic surrounding consisting of orchards and a vineyard.[38]

25. Washburn Observatory, Madison, Wisconsin, c. 1883. (Courtesy of the Mary Lea Shane Archives of the Lick Observatory.)

Having a first-rate observatory, Bascom wanted a first-rate astronomer to lead it. He was convinced Watson was his man. "As he is one of the best astronomers in the United States and has a worldwide reputation, the State University will do well to secure his services," he recommended.[39] Bascom lured Watson with a munificent salary. The observatory at Madison was clearly superior to that at Ann Arbor, whose largest telescope was only a 12-inch refractor. However, the University of Michigan did not let Watson go without a fight. The Wisconsin offer "had the effect to rouse the Ann Arbor people to a more earnest appreciation of Prof. Watson." They offered to match the proposed salary, whereupon Bascom increased Watson's salary by another $500, agreed to pay his then assistant John M. Schaeberle a salary of $500, and appropriated another $350 for mechanical and janitorial services, which Watson had had to provide for himself at Ann Arbor. Bascom was not to be denied. This time the Michigan State Legislature failed to match the offer, and Bascom gloated that he looked forward to "this grand combination—the best instruments and the best man to make the best use of them."[40]

Watson moved to Madison in the summer of 1879. He first took up quarters in Edgewood, Washburn's former home. The site was, however, too far from campus to be convenient. So he moved again, into the president's mansion. He immediately undertook supervision of construction of the east wing of the Washburn Observatory, and with his own money established a Students' Observatory, which somewhat later (under his successor E. S. Holden) was equipped with the 6-inch Clark refractor of the celebrated double-star observer S. W. Burnham.

Vulcan remained his highest priority. As Watson announced to the Board of Regents: "I am . . . erecting, at my own expense, a very elaborate Observatory, one story of which is . . . underground, and one story above ground, which is to be used for special observations of the Sun and the interior planets, and for observations in the vicinity of the Sun. I shall provide the necessary instruments without cost to the University."[41] He was absolutely convinced of the existence of Le Verrier's Vulcan and was certain he had seen that planet. The great aspiration, indeed obsession, of his life was now to prove it conclusively to the rest of the world.

sixteen

Legend into History

Soon after the 1878 eclipse, German astronomer Theodor R. von Oppolzer reworked the chief historical records of possible bodies transiting the Sun: Fritsch (1800 and 1802), Stark (1819), Decuppis (1839), Sidebotham (1849), Ohrt (1857), Lescarbault (1859), and Lummis (1862). Oppolzer published an orbit radically different from any of those proposed by Le Verrier and predicted a transit would take place on March 18, 1879.[1] C. H. F. Peters, as usual, was skeptical and criticized the assumptions underlying Oppolzer's calculations:

> It is incomprehensible that he could claim that his elements satisfy the observations nearly perfectly. . . . Concerning the last mentioned observation, which Le Verrier had called, and Oppolzer also calls, "nearly certain," . . . it proves to be no more than two sunspots, which were observed by me and by Professer Spörer. It is little wonder that the Lescarbault data agrees with the calculations, since they provide the source of the values for the nodes and inclination.[2]

Oppolzer's predicted transit was looked for, but as on prior occasions, the planet failed to register an appearance. Oppolzer entered a brief note in *Astronomische Nachrichten* in which he acknowledged his elements for Vulcan could no longer be regarded as tenable.[3]

The next widely observed eclipse of the Sun occurred January 11, 1880. Conditions were rather unfavorable; totality lasted but half a minute, and at eclipse time the Sun was only 11° above the horizon from the far Pacific coast of the United States, where the eclipse was mainly observed. Immediately afterwards, rumors circulated that the intramercurial planet had reappeared. According to a widely reprinted Associated Press dispatch, "Two professors who observed the total eclipse of the Sun in California, report that they saw an intramercurial planet. This is the Vulcan over whose existence astronomers have had so much controversy. Until the details of the present observation are at hand, those opposed to Le Verrier's theory can at least maintain their opinion."[4] The "professors" were George Davidson of the U.S. Coast Survey and Edgar Frisby of the U.S. Naval Observatory, who had led an eclipse expedition to Mount St. Lucia, California. Peters wrote after talking to Frisby on his return to Washington, "Frisby is just back from the eclipse. They had good weather. . . . The corona was very small; and the Vulcans nowhere. Some fellow in San Francisco, or elsewhere, said that a Vulcan had been found, and we have had many inquiries. How curious and gullible people are." In fact, Frisby and Davidson's results were compromised by cirrus clouds close by the Sun, which prevented them from identifying any bodies at all except Jupiter and Mars.[5] The rumor of Vulcan's having been seen was, according to a contemporary newspaper account, entirely due to the work of "an ambitious or interested reporter."[6]

At the same time Peters wrote of the recurrent "nightmare" of Vulcan, something that had distressed him for 20 years—ever since Le Verrier had put forth his hypothesis.[7] Watson retaliated:

> In view of the fact that I have a personal interest in the question of an intramercurial planet, I must say a few words. . . . Dr. Peters has had the "nightmare" for 20 years past, whenever the existence of such a planet has been discussed. He put the illustrious Le Verrier, one of the greatest astronomers that ever lived, under the ban long ago, because his profound theoretical researches, which had first revealed Neptune, had also proved the existence of another inferior planet. And besides, Dr. Peters, while unquestionably able and industrious as an observer, is inclined to be very crotchety in matters which are the least problematical.[8]

Watson, not surprisingly, regarded the negative observations of Davidson and Frisby as entirely inconclusive, and went on to say:

> I beg to add further that I am one of those who believe in the existence of one or more additional inferior planets. I had the pleasure of seeing one, if not two, of these in July 1878, and I have in my possession conclusive evidence . . . that Mr. Swift, of Rochester, saw also one of these planets, and the one, too, which I certainly recognized as a stranger. . . . As soon as possible I shall communicate to astronomers and to the public further evidence as to the real arrangement of matter near the Sun. Meanwhile, I abandon the field to Peters and his comforters. I shall be consoled by the advice which a distinguished Michigan politician once gave to one of his lieutenants in a political convention, before which his name was pending as a candidate for nomination to a State office. His philosophy was happily expressed in these words: "Now, Sir, while they are making speeches for the other candidates, you peddle my tickets," and he received the nomination.[9]

As he wrote this, Watson was engrossed in building his solar observatory, from which he hoped to obtain the observations of Vulcan that would once and for all refute the arguments of his bitter archrival. The observatory was to be built in an underground tunnel, its plan being based on the idea—no more than a folklorish notion really, but a venerable one, traceable all the way back to Aristotle[10]—that observers who descended into deep shafts and cisterns were able to see the stars in broad daylight. Ironically, the idea had been discredited before Watson tried to put it into practice, but Watson's obsession led him to gloss over the difficulties and to pour several thousand dollars of his own money into the project.

Workmen dug the shaft for the solar observatory into the southward-sloping hill on which the Washburn Observatory had been built (Figure 26). The observatory itself consisted of a cellar $16 \times 20 \times 20$ feet. From the middle of the north wall of this cellar, a pipe 12 inches in diameter and 55-feet long was orientated parallel to the Earth's axis, so that its line of sight pointed directly toward Polaris. The sides of the pipe were reinforced with stone to keep them from caving in. At the top of the pipe was a pier for a clock-driven mirror (heliostat) whose job was to direct the light from the Sun or other celestial bodies down the tube to a 6-inch

26. Washburn Observatory, c. 1891. The small stone structure right of center is Watson's underground solar observatory. (Courtesy of the Mary Lea Shane Archives of the Lick Observatory.)

telescope in the observatory below.[11] In effect, Watson hoped the long shaft would act as an enormous dew cap enabling the observer to sweep close to the Sun's limb and thus espy Vulcan without being blinded by the stray light surrounding it.

Even as this work was getting underway, Watson was putting the finishing touches to his house. While installing a steam furnace, he contracted pneumonia. Unaware of the seriousness of his condition, he did not immediately seek medical attention. He seemed to recover partly, then relapsed, and died unexpectedly on the evening of November 22, 1880. He was only 42.

At his death, the solar observatory had not yet seen first light, and Vulcan remained an enigma. The considerable fortune of $15,000 Watson had acquired through business enterprises during his tenure as director of the University of Michigan Observatory, including actuarial work for the Michigan Mutual Life Insurance Company,[12] was bequeathed not to his widow—a circumstance that raised more than a few eyebrows; she was left nearly destitute[13]—but to the National Academy of Sciences. It was to be invested as a "perpetual fund" for promoting astronomical science. In particular, Watson had asked that "provision be made for preparing and publishing tables of the motion of all the [minor] planets which have been discovered by me as soon as it may be practicable to do so."[14] The bequest was subsequently used by Armin Leuschner of the University of California, Berkeley, to support graduate students who were employed in computing the orbits and ephemerides of the Watson asteroids.

The task of completing the Watson solar observatory now fell to his successor, Edward Singleton Holden. In April 1881, he arrived in Madison from the U. S. Naval Observatory in Washington, D.C. Holden had somehow managed to remain on good terms with both Watson and Peters, and wrote to the latter inviting him to visit at his new residence. Peters replied: "I very readily would follow your invitation, to inspect your establishment out west." However, he was just then about to set sail for Europe. Nevertheless, he could not resist giving Holden some advice: "One thing I would beg you most earnestly: do not sit in that subterranean hole, to watch until Vulcan passes. Not that I apprehend you might discover him . . . but it might deadly ruin your health, and you better fill up the hole, though perhaps objection might be made by the Madison people, with whom Watson seems to have appeared as the greatest human in the world."[15]

Holden eventually completed the solar observatory begun by Watson, but it turned out to be a failure. Watson had expected Vulcan, at magnitude 4.5, to be observable near the Sun from the bottom of his tunnel. Holden found, that even the brightest stars of the Pleiades were invisible at a distance of 50° east of the Sun. In fact no stars were seen at any time. "There is no use in prosecuting this particular experiment further," he concluded. "Every conceivable precaution was taken, and it was shown that this apparatus was not suitable for seeing stars of the magnitude of Vulcan, even distant from the Sun. It would, therefore, be a waste of time to look for stars close to the Sun."[16] Though abandoned for scientific purposes, the stone structure attesting to Watson's tunnel-vision long remained a landmark on campus—it is impossible to verify the rumor that it had once served as living quarters for a poor graduate student. During the 1930s, it was deemed a hazard by the Building and Grounds Department and filled in, or the entrance to it sealed off.[17] It was finally torn down in 1946, and no trace of it remains today.[18]

The next eclipse at which Vulcan was sought was that of May 17, 1882, observed by a party at Sohag, Egypt. As in the case of the 1880 eclipse, totality was brief—only a minute; despite its brevity, it was not unproductive of results. A small comet (unofficially named Tewfik after the ruling monarch of Egypt at the time) was recorded near the Sun. Swift, who after Watson's death became the main torchbearer for the existence of Vulcan, confided to Nashville astronomer E. E. Barnard, "I am half inclined to think it was one of the intramercurial planets."[19] However, photographs obtained at the eclipse clearly showed the object's cometary nature. The comet vanished and was never seen again.

The next opportunity to search for Vulcan was the May 6, 1883, eclipse, visible from the central Pacific. Swift himself hoped to observe it, and wrote to Barnard he "ought to go to verify my previous discovery of the 2 intramercurial planets."[20] Unfortunately, he was unable to do so, but Holden led an American team to Caroline Island, a thinly inhabited tiny atoll in the central Pacific (not, as the name might suggest, in the Caroline Islands but in the group called the Line Islands, located about a quarter of the way from Tahiti toward the Hawaiian Islands). Within a few days they were joined by a French team under the direction of Jules Janssen of the Meudon Observatory (Figure 27).

After the eclipse, there were rumors of "a very red star" seen in the finder of one of the telescopes some $3°$ northwest of the Sun by a member of the French team, E. L. Trouvelot, and his assistant. Unfortunately, they had to switch to preassigned duties before they could confirm its position. Although Swift concluded Trouvelot must certainly have seen the planet, more definitive searches carried out at the same eclipse yielded nothing.[21] Holden used the 6-inch Clark refractor the University of Wisconsin had recently acquired from double-star observer S. W. Burnham to search for intramercurial planets. He was able to satisfy himself that "no star as bright as 5 ½ magnitude could have escaped me." Johann Palisa of the Vienna Observatory, who later became an even more successful visual searcher for asteroids than Watson or Peters, also searched for the intramercurial planet, with the same want of success. Holden concluded: "It is my opinion, therefore, that at future eclipses it will not be necessary to devote an observer and a telescope to the further prosecution of this search, and I must regard the fact of the non-existence of Vulcan as definitively settled by Dr. Palisa's observations and my own."[22]

The British amateur astronomer T. W. Backhouse and several of his colleagues monitored the sun during the spring and fall nodes of 1884–1885, again without success. "By 1886," writes Vulcan historian R. Fontenrose, "Vulcan's stock had reached what was to be a permanent low."[23] A statue of Le Verrier (Frontispiece) was erected in 1888 in the north court of the Paris Observatory. It was paid for by subscription to which Watson had generously contributed and contained on its pedestal a representation of the solar system, which included Vulcan. However, Vulcan was subsequently rubbed out. This only acknowledged the growing consensus. Peters, before his sudden death, had the satisfaction of seeing

27. Caroline Island Eclipse Expedition, May 1883. Simon Newcomb seated, arms folded, second from right; Edward S. Holden seated, checkered shirt, third from right. (Courtesy of the Mary Lea Shane Archives of the Lick Observatory.)

his position fully vindicated; "he was happy because he had stood fast in refusing to accept a conclusion drawn from incomplete and imperfect data."[24] On July 18, 1890, he collapsed, a half-burned cigar in hand and observing cap on his head, on the steps of the college building where he lived, on his way home after a night's observing with the refractor of the Litchfield Observatory. Other than a few eccentrics of which nothing need be said, almost the lone holdout was Swift. Still committed to his observations of July 1878, he clung steadfastly, and increasingly desperately, to the reality of the intramercurial planets.[25]

And yet the problem that Vulcan (or the Vulcans) had been invoked to explain remained as intractable as ever. The anomalous advance of the perihelion of Mercury was unsolved. Indeed, even in its heyday, Vulcan had never provided more than a fraction of the missing mass needed to satisfy the 38 arc seconds per century by which Le Verrier had found Mercury to be precessing ahead of schedule. According to Newtonian theory, the mass required to produce such a precession must be equal to that of Mercury itself. This would have required the existence of millions of objects the size of Watson and Swift's "Vulcan." Richard Proctor summed up:

> Whatsoever matter must be assumed to travel within the orbit of Mercury, to account for the motion of the planet's perihelion, is evidently neither gathered into a single planet nor distributed among several bodies which, though small, could be regarded nevertheless as planets.... The only supposition which remains available is, then, that the matter within the orbit of Mercury consists of multitudinous small bodies individually invisible. Many among these may be several tons, or hundreds of tons, in mass; but (when considered with reference to the enormous region they occupy, and compared with the masses of even the smallest planets) they must be regarded collectively as mere planetary dust.[26]

Theoreticians were generally baffled. Considering the problem in 1882, Simon Newcomb summed up that Le Verrier's hypothesis was no longer tenable, and although there was no way of "positively disproving" the asteroidal ring hypothesis, the quantity of matter required "would glow with a much brighter light than the zodiacal light." Moreover, the asteroidal ring must have a high inclination in order to explain the advance of the perihelion of Mercury without at the same time producing a larger-than-observed regression of Mercury's and Venus's nodes.[27] At the time, Newcomb was not quite sure what to suggest.

By 1895, the problem had become even more acute. New-comb that year published tables for the four inner planets (Mercury, Venus, Earth, and Mars), on which he had been working ever since 1877, the year of Le Verrier's death. It had been a massive undertaking and was necessitated in part by the fact that to make the theory of each planet fit the observations, Le Verrier had attributed different masses for one and the same planet in each theory in which it figured as a perturbing body. This was an obvious inconsistency, and one in Newcomb's view "injurious to the precision and symmetry of much of our astronomical work." He saw no alternative but to start over. In the course of his reanalysis, Newcomb found the mass of Mercury had to be reduced by half and that the anomalous advance of its perihelion was even greater than Le Verrier had supposed—43 rather than 38 arc seconds per century.[28] He considered several alternatives. It might, he suggested, be due to a nonspherical distribution of matter within the Sun itself. However, there was no evidence the Sun had the required oblate shape; to the contrary, the observations pointed the other way. The idea of a highly inclined asteroid ring was still in contention. Still, there was the obvious problem of how, if such a ring existed, it could have escaped detection. Indeed, hidden mass theories of the sort favored by Le Verrier were now generally déclassé. The confidence the problem could be solved from within the settled framework of Newtonian theory had all but completely lapsed. *"Le roi Le Verrier, l'était mort,"* wrote Norwood Russell Hanson; "his system of celestial government lay moribund; a host of rival hypotheses quickened the dissension initiated by Mercury, finding flaws in other parts of Newton's plan."[29]

On the whole, Newcomb at this time favored the theory offered by his long-time colleague at the U.S. Naval Observatory, Asaph Hall (Figure 28), who had set to tinkering with the inverse-square law of gravitation itself. Rather than the inverse-square law, Hall decided, by means of an ad hoc calculation, the exponent ought to be 2.00000016.[30] The modification was minor, its

Asaph Hall.
(Photo. by Chickering, Boston.)

28. Asaph Hall (1829–1907). (Courtesy of the Yerkes Observatory.)

implications vast. In effect, it was calculated to define an intramercurial planet out of existence. "The conceptual shift," added Hanson, "is of historical significance. Once one is prepared to modify the foundations of Newtonian theory in order to accommodate the facts, the possibility of rejecting the *whole* theory—replacing its foundations—becomes genuinely a live option. Every line of Le Verrier's work reveals that he could never have done this. Everything else would have been challenged, anything else would have been hunted, to preserve the Newtonian theory intact. But the theory's failure with Mercury quickened the pace of ideas and the pace of conceptual adjustments."[31]

By 1896, Swift had fallen on hard times. The Warner Observatory in Rochester, New York, where he had spent his most productive years, had closed after its patron lost his fortune in the financial panic of 1893, and Swift had departed for Thaddeus Lowe's observatory at Echo Mountain, California. He now published a lonely defense of the intramercurial planets. His remarks appeared in the *English Mechanic*. There the intramercurial planet observations had been treated derisively by a regular correspondent, "FRAS" (pseudonym of Captain William Noble, first president of the British Astronomical Association, 1890–1892), who advised that further searches would be a complete waste of time. Swift briefly recalled his observations of the strange "red stars," which he had seen at the 1878 eclipse. Alluding to Peters's hypothesis that the objects he and Watson had seen were none other than the well-known stars θ and ζ Cancri, he argued, "last evening (Aug. 31), I examined α^1 and α^2 Capricorni, and, to my mind, they resemble, as to distance apart and to brightness, the stars seen at Denver. That one of my stars was Theta Cancri and the other Zeta is entirely absurd, as the interval between these exceeds 4° 45', and the line joining them points not to the Sun's center, but far from it."[32] Obviously Peters's criticisms still rankled. Swift, and Watson, too, for that matter, had taken them rather personally—on one occasion, Swift had even confided to E. E. Barnard that Peters "without a semblance of a cause hates me."[33] Swift objected to Peters's "futile attempt to criticise Prof. Watson's observations, but the latter as an astronomer, both mathematical and observational, was Dr. Peters's superior, and the treatment which the last-named gentleman accorded to Watson's observations was most discreditable."[34]

And yet Swift was forced, after all, to admit that he and Watson had recorded different objects. Later he said: "I am not a believer in the Vulcan of Lescarbault—perhaps I do not impugn his honesty, but think he saw, perhaps a small round sunspot regarding whose motion he was simply mistaken." These concessions did little to strengthen the waning case for Vulcan. Nevertheless, he still hoped:

> This lack of faith does not militate against the existence of a swarm of intramercurial planetoids which, for all anyone's knowledge to the contrary, may be numbered by thousands. . . . Let us, in

closing, recall the advice of his wife to Professor Hall to be patient in continuance in his search for a satellite to Mars; that of Le Verrier to Dr. Galle to seek in a certain region for a new planet; the division of the sky into zones, each assigned to a different astronomer for the finding of an invisible planet between Mars and Jupiter, with what success in every case the world well knows, and contrast these expectant and hopeful urgings with the pessimistic advice of "FRAS" to waste no time in the search for intramercurial planets.[35]

What of Lescarbault? Largely unheard of since he had received the Legion of Honor and continuing to practice medicine in obscure Orgères, this generally private little man ventured, on January 11, 1891, another astronomical "discovery" to the Secretary of the Académie des Sciences: He reported seeing "a star of comparable brightness to Regulus, which I had never seen until today. It is beneath Theta in the Lion. I observed it only with the naked eye, on 10 and 11 January, in the early morning hours; and despite the weakening of my eyesight, I believe I saw it well, and was not the victim of an illusion."[36] His (and Le Verrier's) longtime critic Camille Flammarion hastened to point out, with what satisfaction one can only imagine, that the star was none other than the well-known planet Saturn![37] Only death, which came to Lescarbault in 1894, could rescue him from such a red-faced embarrassment. In the process of "discovering" Saturn he had inevitably, and irreparably, damaged the credibility of his earlier observation.

"FRAS," commenting on Swift's essay in the *English Mechanic*, put a point to the obvious: "I am sincerely glad to find my utter disbelief in the existence of 'Vulcan' shared by so competent an authority as Dr. Swift. When we know that Dr. Lescarbault 'discovered' Saturn! and communicated his discovery to the French Academy of Sciences, I think that any unprejudiced person will agree with me that any astronomical observation that he may claim to have made is not worth the breath it was described with or the paper on which an account of it was written."[38] Harsh words indeed. But we must remember Lescarbault's age; he was then in his 77th year, his faculties failing. He made a mistake and that is all.

Peter Lancaster Brown in his *Comets, Meteorites and Men* tells of one modern observer well known for his discoveries who alerted a colleague to a possible new comet. Upon close examination the "comet" proved to be no more than a lump of damp snow on a telephone wire dimly lit by a street light! Another experienced contemporary, more youthful than Lescarbault, was also fooled by a light ghost from a nearby TV mast into believing he had discovered a new comet. There are many such instances registered in the history of observational astronomy.[39]

Vulcan, or its congeners, continued to be scouted on future occasions. Swift, on September 20, 1896, saw a peculiar luminous object above the Sun at sunset over the Sierra Madres. "Seizing an opera-glass I saw that it had a very much fainter companion some 30′ north," he wrote, "but it could not be seen without the glass. In about four minutes after the sun had set, the two objects also disappeared behind the mountain."[40] To Swift, both objects had a "hazy, cometary appearance." He later suggested they might well have been comets. The following evening Swift saw a single faint object near the setting Sun.[41]

Two observers noted a bright body "certainly not a star" midway between Venus and Mars at the eclipse of January 22, 1898. It was only seen briefly before totality, and was estimated as of the 2nd magnitude.[42]

Such sightings did little to revive interest in Vulcan. In 1899, Asaph Hall concluded the intramercurial planets had disappeared from "rational astronomy." "I think they are not mentioned even by the astronomical writers of the *Atlantic Monthly*."[43] A year later the leading British amateur and meteor expert W. F. Denning spoke for all visual observers when he said:

I have obtained some thousands of solar observations with different instruments, but chiefly with refractors of 4¼ inches and 3 inches, and a reflector of 4 inches aperture with a view to the detection of an intramercurial planet. The months of observation were usually March–April, and September–October. On some days the Sun was examined at short intervals during the whole time that he remained above the horizon, but I never met with any object representing an intramercurial planet. Occasionally a suspicious looking spot— pretty round, small and without penumbra—was noticed, but upon being closely watched it always proved a veritable sun-spot. I believe that spots with certain planetary aspects appear more often than is generally supposed, and perhaps it is no wonder that their character has been sometimes mistaken by persons who have formed hurried conclusions without applying proper tests.[44]

Photography had advanced quickly with the introduction of the dry-plate and by 1900 had completely supplanted the desperately hurried and unreliable searches of visual observers like Denning, Swift, and Watson. At the May 28, 1900, solar eclipse, Harvard's William Henry Pickering, fresh from the triumph of making the first photographic discovery of a satellite, Saturn's Phoebe, a year before, searched inconclusively for Vulcan with a specially designed wide-angle lens.[45] Having photographed an extensive area around the eclipsed sun, C. G. Abbot of the Smithsonian Astrophysical Observatory reported, "the absence of intramercurial planets above the fourth magnitude made nearly certain, and the presence of several such between the fifth and seventh magnitude rendered as probable as single photographs can do."[46]

At the eclipse of May 18, 1901, Abbot and Perrine, in Sumatra (now Indonesia), the former with the Smithsonian Expedition, the latter in charge of the Crocker-Lick Expedition, planned to take duplicate wide field photographs covering a large area east and west of the sun, so they prepared themselves with four cameras each. Conditions were against Abbot, but thin cloud at the critical time allowed 25% of the light through to Perrine's plates. He imaged an area $6° \times 38°$ extending along the direction of the Sun's equator, with the Sun located centrally. He recorded 170 known stars, with stars down to magnitude 8 and fainter being imaged in two-thirds of the area covered. In the remaining third, which was affected by thicker clouds, the limiting magnitude of the stars recorded was 5 or 6.

To observe the eclipse of August 30, 1905, temporary observing stations in Labrador, Spain, and Egypt were set up by the Lick Observatory. Each was equipped with four special wide-angle cameras, designed and constructed under the supervision of Perrine. Severe storm conditions disabled the Labrador party. Hussey in Egypt had more success: He recorded a considerable number of stars, but airborne dust and the low altitude of the Sun impaired his results. Thin cloud allowed about 20–30% of the light through to Perrine's cameras in Spain. He registered 55 stars down to about magnitude 8; several suspect images were later identified as film defects.

Cloud and rain interfered at the start of totality in 1908, but clear sky prevailed in the last 160 seconds of the four critical minutes. Perrine got duplicate exposures east and west of the Sun, and counted over 500 stellar images down to visual magnitude 9 on the plates, but no unknown planets. Henceforth even the existence of large meteoritic bodies was seriously in doubt, since Perrine's survey had the capability of revealing objects down to 30 miles in diameter.[47] Millions of such bodies were needed to furnish enough mass to explain the anomalous advance of Mercury's motion.

In 1909, W. W. Campbell said he believed none of the reported instances of intramercurial planets had any foundation in fact. Watson and Swift's observations were still the best candidates; however, as veteran of several eclipse expeditions, Campbell knew only too well the unique madness that so often prevails at totality and shared Peters's assessment of what had happened. Hence, "as the assigned locations depended upon the hasty readings of graduated circles, in which one can so easily make errors, in the press and excitement of eclipse conditions, the astronomical world quickly, and no doubt correctly, concluded that the objects seen were well-known neighboring stars." For the previous 20 years the Sun had been photographed every clear day, but no transit of Vulcan or any other intramercurial body for that matter, had been authenticated. Indeed as limiting magnitudes got progressively lower, the quarry became increasingly more elusive. "Taking all these points into consideration," Campbell wrote in *Popular Science Monthly*, "I think we may say that the investigations by Perrine, forming a part of the work of the Crocker Eclipse Expeditions from Lick Observatory, have brought the observational side of the intramercurial problem, famous for half a century, definitely to a close."[48] By then, the whole business had come to seem rather like flogging a dead horse.

If a chapter of the observational problem was closed by Campbell's 1909 paper, the theoretical problem of the anomalous advance of the perihelion of Mercury remained open. After the turn of the century, the German astronomer Hugo von Seeliger's "zodiacal light hypothesis" enjoyed a brief heyday. Seeliger had inferred the existence of ellipsoidal concentrations of small particles around the Sun. One such ellipsoid was concentrated so near the Sun as to be invisible. This was supposed to be responsible for perturbing the motion of Mercury. Another ellipsoid, extending just beyond the orbit of the Earth, was observable as the zodiacal light. The eminent French mathematician Henri Poincaré inclined to the "circumsolar ring" hypothesis, and Simon Newcomb, too, before his death in 1907, was also leaning strongly in that direction—in part because it seemed (incorrectly, as it turned out) that Hall's revision of the inverse-square law led to unacceptable consequences in the motion of the Moon. However, it was objected that in the circumsolar ring hypothesis, an observed perturbation of comets could be anticipated. Moreover, it was not entirely plausible that an ellipsoid sufficiently massive to perturb Mercury's motion would remain completely invisible. The matter remained problematic.

In the end, the astronomers never did solve the problem; instead its resolution came from radical new ideas in physics. The correct, non-Newtonian explanation of the discrepancy in Mercury's motion was found by Albert Einstein (Figure 29) in November 1915 at a desk in Berlin—the city where only 70 years earlier the discovery of Neptune had brought Newtonian theory to its bright zenith.

Einstein, at 35, was the same age as Le Verrier at the time of Neptune's discovery. Unlike the Frenchman, his work was not at all aimed at solving the anomalous advance of Mercury's perihelion. He was in fact looking for a general complete physical theory; his solution to the problem fell out automatically. He later wrote it was clear that the anomaly could be explained by means of classical mechanics "only on the assumption of hypotheses which have little probability, and which were devised solely for this purpose."[49] In 1915 he had just introduced new gravitational equations derived from his General Theory of Relativity. Simply put, he proposed that in the neighborhood of small bodies such as

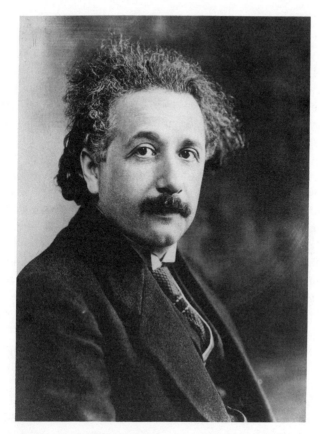

29. Albert Einstein (1879–1955). (Courtesy of the Yerkes Observatory.)

the Earth, the relativistic adjustment to Newton's law is negligible, but close to a massive body such as the Sun, there is a significant curvature of Einsteinian space-time, which produces a non-Newtonian warp in the trajectory of nearby bodies such as Mercury. Einstein found that according to his theory of gravitation, Mercury should precess slightly faster than the Newtonian rate. In fact, the excess comes out to 0.1 arc seconds for each orbital revolution of the planet, or 43 arc seconds per century; this calculated value was, of course, identical to the 43 arc seconds found by Newcomb.[50] After making this calculation, Einstein later confessed, he was for several days beside himself with excitement.[51]

Einstein's calculation of the advance of the perihelion of Mercury was one of the most dramatic early triumphs of General Relativity. With its publication begins the post-Newtonian era of celestial mechanics. The case of Mercury is perhaps the most satisfying empirical proof of the correctness of Einstein's ideas.[52] However, since 1975 a more dramatic case has been studied: the binary pulsar PSR 1913+16, which consists of a pair of massive stars orbiting around one another in an elliptical orbit, with a minimum separation about equal to the Sun's radius. One of the stars is a collapsed star known as a pulsar and emits regular radio pulses, allowing its position and velocity to be determined with extraordinarily high accuracy. The perihelion has been shown to advance strictly in accordance with Einstein's theory, at a rate of 4.23° *per year!*[53]

General relativity had at last abolished the need for Vulcan (or Vulcans), indeed for any significant amount of circumsolar matter. The ghost had been exorcized, or nearly so, and the Abbé Moigno's "planet of romance" became, as Richard Proctor said, "the planet of fiction." Like the Martian "canals" of the 1890s, the intramercurial planet was an illusion, an all-too-compelling chimera. It was created largely out of wishful thinking, the need to avert disaster for Newtonian theory.

And yet questions remained. What, if anything, did Lescarbault see on March 26, 1859? Probably he was simply mistaken. Liais's failure to see anything while looking at almost the same moment was a blow from which Lescarbault's observation never recovered. Whatever he saw, it was certainly not intramercurial. Instead, as Liais pointed out, his object might have been located anywhere on a line between the Earth and the Sun, and if actually closer to the Earth than Lescarbault (and Le Verrier) assumed, could have shown a parallax large enough for it to have been on the Sun's disk at Orgères, but not at San Domingo.

The case for Watson and Swift is more compelling. Unlike Lescarbault, they were skillful observers and absolutely convinced of what they had seen. In 1878, Vulcan was still very much in the air, and it was natural they should conclude in favor of their expectations. That they saw something near the Sun is certain. Peters's misidentification hypothesis is not as conclusive as it once seemed. In particular, as Heber D. Curtis, the distinguished University of Michigan astronomer, noted, "Watson's observations have never been explained in a manner that is entirely satisfactory. It must be remembered, in assembling the available evidence, that, practically all Watson's working life had been spent in making and studying star charts, and in detecting new objects which were often at the limit of visibility. It seems inconceivable an astronomer of such wide experience in this particular field could have been deceived or made a mistake."[54]

If he did not misidentify known stars, what did he see? Possibly a member of the Earth-crossing or Apollo group of asteroids with an orbit that takes it inside the orbit of Mercury, as is the case with Icarus, discovered in 1949 by Walter Baade. Or as Curtis suggested "a moderately bright comet with an almost stellar nucleus; the relatively strong illumination of the sky background would have prevented the detection of a tail."[55] After all, comets do brighten enormously as they approach the Sun.

In recent years numerous examples of small sungrazing comets (pygmy sungrazers), such as Comet Tewfik of 1882, have been recorded by the Earth-orbiting Solar Max and SOLWIND satellites. Some have actually collided with the Sun. There may even be a continuous stream of such objects, as William Thomson (Lord Kelvin) conjectured in 1854. That Watson and Swift saw one or more of these pygmy comets near the Sun at the eclipse of July 28, 1878, is an appealing explanation. Of course, we shall never know for certain.

Whatever mystery remains, little doubt exists that in retrospect 1878 was the watershed, the division between what had been and what was to come. Vulcan had made its last stand. First Separation, Wyoming, disappeared when Union Pacific moved their track to the south. On July 12, 1893, a young professor of history, Frederick Jackson Turner, addressed the American Historical Association in Chicago on the closure of the American Frontier. A decade or so later, Lick observatory director William Wallace Campbell proclaimed "The Closing of A Famous Astronomical Problem." Vulcan, too, had finally passed from legend into history.

V

LAGRANGE'S
LONG SHADOW

seventeen

A Question
of Triangles

When W. W. Campbell published "The Closing of a Famous Astronomical Problem" in 1909, he overlooked an interesting possibility. On February 22, 1906, Max Wolf at Königstuhl photographically discovered a 12th-magnitude object. Provisionally designated 1906 TG (subsequently named Achilles), it had a mean daily motion of 300 arc seconds and moved in a nearly circular orbit.[1] Impossible! There must be a mistake. That meant it traveled in the orbit of Jupiter. How could such a tiny body resist the gravitational pull of the most massive of planets? The answer emerged when C. V. L. Charlier of Lund, Sweden, found the object was 55° ahead of Jupiter; this, he pointed out, corresponds to the Langrangian solution of the classic three-body problem.

As is well known, the dynamics of the two-body problem are not difficult to describe; bodies move along fixed and relatively simple curves in accordance with Kepler's laws of planetary motion, and the configuration is stable. With the three-body problem, however ("the despair of all astronomical computers"), orbits, generally speaking, are neither closed curves, nor do the bodies follow the same path in any two circuits about the center of gravity.

Nevertheless, in 1772, Lagrange showed that in one special case the gravitational stability exemplified by one body revolving about another is also to be found in the case of three bodies, provided one of the three is of neglible size. In fact there are five Lagrangian points. Three are called the *straightline solutions* and are unstable to small perturbations; the other two, the *equilaterals*, are stable (Figure 30).

Taking Jupiter as an example, the *straightline* points L_1, L_2, and L_3 are all on the line that connects the planet and the Sun. The L_3 point is on the opposite side of the Sun to Jupiter, the L_1 point is between the Sun and Jupiter, while L_2 lies beyond the planet. The L_4 and L_5 points are located in the plane of the planet's orbit 60° ahead

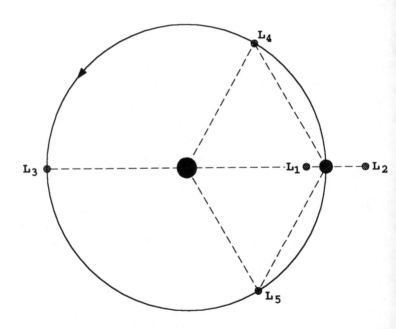

30. Lagrangian points. Points in the orbital plane of two massive bodies in circular motion about their center of gravity, where bodies of neglible mass can remain in equilibrium. The L_1, L_2, and L_3 are the straightline solutions. L_4 and L_5 are the equilateral solutions.

and 60° following Jupiter, and with the planet and the Sun form two *vast equilateral triangles*. That Achilles was 55° distant from Jupiter instead of 60° was accounted for by the gravitational disturbances of the other planets, which produce slight swinging motions, causing objects to perform complicated oscillations about the equilateral points. Eventually other asteroids were found at the L_4 position, while another group was discovered at L_5, i.e., 60° trailing Jupiter. All these small bodies were named for heroes of the Trojan War, hence the group name, the Trojan asteroids.

The discovery of the first Trojan asteroids, fulfilling as it did Lagrange's elegant concept, induced Charlier to ask, if Trojan asteroids, why not Trojan Vulcans? He believed Mercury was formed by a process of accretion; small bodies gradually coalesced under the influence of mutual gravitational attraction. If such was the case, he suggested in 1912, was it not possible to expect discrete accumulations of residual material at the planet's L_4 and L_5 points?[2] H. Block, also of Lund, established the possibility by demonstrating that the stability of Mercury's triangle points was unaffected by the high eccentricity of the planet's orbit.[3] Charlier tested the hypothesis at the 1914 eclipse but his plates of one triangle point were densely fogged.

By then Europe was at war and further investigation was impossible. In 1921 Charlier revived his interest. Two years later in a move reminiscent of Le Verrier and Galle, he solicited the help of Lick Observatory's Robert Trumpler. Like Galle and Neptune, Trumpler favored the idea and made arrangements to observe the 1923 eclipse from Ensenada, Mexico, with the 5-foot cameras previously used by Perrine.[4] Unfortunately the opportunity was lost to clouds. As a check, Trumpler reexamined plates from the previous Lick eclipse expeditions, the intention being to see which of the triangle points had been covered; only those taken in 1901 included both positions. Even so he thought the results spoke "decisively against the existence in these regions of small planets brighter than the 8th photographic magnitude." As such small planets probably would have considerable librations around the triangle points, "the negative results of the searches of 1908 and 1922 extending close to L_4, also have some bearing on the problem."[5]

It is worth pointing out that objects near the triangle points would be more readily observable than intramercurial planets, as they are visible at a much larger elongation distance (the same as Mercury, 28°). Under favorable conditions, they would register photographically at times other than total eclipses of the Sun. Moreover they ought to transit the Sun less often than the intramercurial planets predicted by Le Verrier. Trumpler decided to search for them, "it is necessary to make the observations at the period of shortest dawn (for Mount Hamilton around Oct. 6 and March 9) and about between 4h and 8h sidereal time when the plane of the ecliptic is most nearly vertical." It is possible, he added, "to make such observations for western elongations (to be observed shortly before sunrise) between the middle of September and the end of October."[6]

Such an opportunity occurred at the end of October 1923, when L_5 was favorably placed. Photographs were obtained with two 6-inch cameras (33-inch and 31.5-inch focal length) on the mornings of October 28 and 31, and again on those of November 2 and 3. The total area of the search covered a rectangle 6° wide extending 4° to the east of L_5 and almost 15° to the west. A clear photographic signature could be expected of a small planet in this position having the mean motion of Mercury. "No trace of any small planet was found," and Trumpler concluded "no such bodies brighter than the photographic magnitude 11.0 exist near the triangle point L_5. At the mean epoch of the search a small planet of the 11th magnitude . . . situated at L_5 would be 40,000 times (11.6 magnitudes) less luminous than Mercury in the same position would be. Even if such a planet had the small albedo of Mercury and the Moon, its diameter would be only 14 miles, and its volume 8 million times less than that of Mercury. In a similar way we find that at the 1901 eclipse a planet of the photographic magnitude of 8.2, situated at L_4 would be 4900 times (9.8 magnitudes) fainter than *Mercury* in the same position; or its diameter would be 39 miles."[7]

Meanwhile, not yet yielding to the persuasiveness of Einsteinian views, British astronomer W. M. Smart announced a theoretical investigation into the Trojan solution in 1921 as "an exten-

sion or completion of Le Verrier's calculations." Six Trojan asteroids had then been discovered, and Smart justified his study in the
knowledge that "the equilateral configuration seems to be a definite feature of the solar system."[8]

Smart reasoned the mass of a disturbing body necessary to account for the observed rate of advance of Mercury's perihelion
ranged from ⅙ to 2⅗ that of Mercury for heliocentric distances of
0.27 to 0.12 AU, the heliocentric distance of Mercury being 0.39
AU. In other words, this suggested a considerable stellar magnitude for Vulcan, which almost certainly would have led to discovery during a total eclipse of the Sun.

The large gravitational disturbances resulting in the equilateral three-body problem, Smart argued, suggest "the possible existence of a Vulcan of small mass at the same heliocentric distance
as Mercury." He determined then to find the mass of this hypothetical body so that the anomalous advance of Mercury's perihelion may be accounted for in the Newtonian law and also to obtain a rough approximation of its stellar magnitude that would
"afford some indication of the possibility of its discovery (or
not)." If its mass was much less than that of Mercury, the difficulty of establishing its existence observationally would be great;
if comparable with Mercury, however, Einstein's critical test
could not be challenged.[9]

He found a Trojan Vulcan of ½ the mass of Mercury, with an
estimated maximum stellar magnitude between 0.0 to 1.2, was
necessary to account for the well-known discrepancy. "Vulcan has
been searched for at total solar eclipses," he concluded, "and continuous observation of the Sun has failed to detect the transit of the
hypothetical planet across the Sun's disc. The probability of the
planet's existence seems from the foregoing to be very small." But
he continued if "the mass of the Trojan Vulcan had come out at
1/700 or even 1/70 mass of Mercury, the circumstances of the
planet's non-discovery would have been easily explicable on observational grounds.[10]

In reference to the controversial Watson–Swift objects Smart
noted, "for a Vulcan preceding Mercury, the solar eclipse of 1878
would not have proved a favourable occasion for the detection of

the planet, when the area swept is taken into consideration, although a star of magnitude 0.0 to 1.2 at a distance from the Sun of 12°–23° might have been expected to attract attention."[11] Trumpler concurred. Summing up in November 1923 he said: "We may say that the observational evidence is against the existence near the triangle point L_5 of planets larger than 14 miles in diameter, and near L_4 of planets larger than 39 miles in diameter. Bodies near Mercury's triangle points smaller than the given limit cannot be the cause of the motion of the perihelion of Mercury. In order to make up sufficient mass ($\frac{1}{2}$ of the mass according to Smart) they would have to be so numerous as to make the two regions visible like diffused nebulous objects."[12]

The intramercurial planet problem had clearly changed. Systematic searches had left little probability that such bodies existed unless they were exceedingly small and faint. In the meantime Campbell wrote,

> the problem of the motion of the perihelion of Mercury which had been the starting point for the search of an intramercurial planet, changed its aspect. Seeliger's hypothesis on the structure of the Zodiacal Light and more recently Einstein's Generalized Theory of Relativity offer explanations which make the assumption of one or several perturbing planets inside of Mercury's orbit unnecessary. Although these developments may have reduced its importance, the question, whether there are any small planets closer to the Sun than Mercury, has an interest of its own; and it is desirable to extend the search for such bodies to the utmost limit of our instrumental means.[13]

Accordingly a plate Campbell had taken at Wallal, Western Australia (Lick-Crocker Expedition), during the total eclipse of September 21, 1922, to measure the light deflection predicted by Einstein's Theory of Relativity, was closely examined by Trumpler. It contained the images of over 550 stars down to photographic magnitude 10.2 and covered an area 15° square with the eclipsed Sun in the center. Chapman and Melotte believed an area 225° square at that galactic latitude should, on average, contain 830 stars brighter than photographic magnitude 10.0. "Although the difference is well explained if our field is poorer in stars than the average," observed Campbell and Trumpler, "it is possible that the search is not complete between magnitudes 9.5 and 10.2, especially within the limits of the corona, and near the corners of the plate where the star images are somewhat less sharp."[14] Still it was obvious "a planet brighter than photographic magnitude 9.5 within the field of the plate could hardly have escaped detection unless it was situated in the denser part of the corona or had a sensible apparent motion during the one minute exposure."[15]

Plates exposed at the Sumatra eclipse of 1929 by Einstein's friend and colleague Erwin Freundlich, director of the Einstein Tower, Potsdam, showed such a profusion of star images that H. von Klüber decided to test the Vulcan hypothesis. Six months later, he photographed the region with the same telescope from Potsdam, i.e., when the Sun was in the opposite part of the sky. The comparison proved conclusively there was no planet as bright as the 9th magnitude up to a distance of 40 arc minutes (approximately equal to 1⅓ solar diameters) from the Sun. Fainter objects much closer to the Sun would almost certainly have been rendered invisible by the brightness of the corona. Even then von Klüber believed it would be possible to pick out a planet as bright as the 7th magnitude. However no suspects were noted, and he decided there can be no significant body between Mercury and the Sun.[16]

Even so, the ghost of Vulcan continued to haunt the minds of some astronomers, who for whatever reason found Einstein's ideas disquieting. Others pursued it either because they hoped to find in it the missing matter needed to fulfill theory or for its own sake—because they felt the solar system might not yet be completely inventoried. So the quest for Vulcan, even after the triumph of Einstein's theory, continued well into the 20th century. Indeed there was a faint stirring of Buys-Ballot in E. Huntington's 1923 work on the possible planetary influence on variations in sunspot numbers and weather phenomena.[17]

On June 19, 1971, at an international symposium in Seattle, Washington, Henry C. Courten reported on nonstellar objects imaged during the total solar eclipses of 1966 and 1970. Significantly, orbital eccentricities derived for three of these indicated they were apparently in heliocentric orbit about 0.1 AU from the Sun.[18]

In 1973, F. Dossin and A. Heck of the Astrophysical Institute of Leige, Belgium, paralleled the eclipse observations of 1869, 1878, and 1898 with a report of an unknown object (magnitude −2.0) close to the Sun at the June 30 total eclipse. The object was featured on 20 plates taken with two cameras stationed at Loyiengalani, Kenya.[19]

Again, in 1979 Tisserand's advice "to return to the idea first given by Le Verrier, that there is a ring of asteroids between Mercury and the Sun,"[20] was recalled in the science weekly *Nature* when Kenneth Brecher and others queried: "Is there a ring around the Sun?" Their question was based on the fact that as rings are now known to exist around the four major planets, why not around the Sun itself? Although the planetary system already forms such a ring, they envisioned a ring of rocky material lying a few solar radii from the Sun. A minimum particle size of 20 km was predicted. Such bodies might be individually detectable if they have a visual magnitude of +8 to +10 during a total solar eclipse. Some infrared data hints at the existence of such a ring.[21] Thus the possibility of tiny intramercurial bodies remains; the spell of astronomy's most famous revenant has not yet been broken.

When we recollect the solar system is webbed by a complex tracery of intersecting orbits mapped out by myriads of tiny bodies, boulders, extinct comet nuclei, and dust particles, we also acknowledge the possibility that part of this material might on occasion pass in front of stars, planets, satellites, or the Sun. In the case of point light sources, i.e., stars, an occultation would be observed. Extended bright objects like the Sun, Moon, or planets would act as a screen on which the object would show up as a moving dark spot.[22] Could Lescarbault and others have witnessed such a passage?

If the standard explanations are disconnected and the observational data examined afresh, we inevitably discover possibilities not at all obvious to our predecessors. Add to this suspicions of an enrichment of small bodies in the inner solar system in astronomically recent times, and suddenly the rejected Vulcan data assumes a new significance; "1,500 foot asteroid comes close to making the Earth move" proclaimed a headline on May 21, 1996. A timely, if chilling reminder of the Great Jupiter Crash of 1994.

The Unending Quest

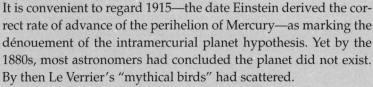

It is convenient to regard 1915—the date Einstein derived the correct rate of advance of the perihelion of Mercury—as marking the dénouement of the intramercurial planet hypothesis. Yet by the 1880s, most astronomers had concluded the planet did not exist. By then Le Verrier's "mythical birds" had scattered.

Nevertheless, the grand triumph of the discovery of Neptune by Le Verrier and Adams continued to exert a powerful spell. The dream of emulating their immortal success by revealing other worlds lured a select band of astronomers who became devoted planet-seekers during the century after the glow of Vulcan faded.

With Vulcan's failure to materialize, the attention of these planet-seekers ranged from the inner solar system to its outer fringes, the yawning gulf beyond Neptune. As earlier mentioned, Watson himself had contemplated a survey of transneptunian space aided by his special charts of the stars along the ecliptic, but the Vulcan detour and his premature death prevented his quest. David Peck Todd made a brief foray into the field at about the same time. Based on a graphical analysis of the residuals remaining in Le Verrier's 1873 theory of Uranus, he deduced the existence of a possible planet at a distance from the Sun of 52 AU and went so far as to look for it with the 26-inch refractor at the U.S. Naval Observatory. However, the host of background stars proved unyielding.

E. S. Holden, Todd's onetime colleague at the U. S. Naval Observatory and Watson's successor as director of the Washburn Observatory, also meditated a search for a transneptunian planet. In 1888, shortly before he assumed the directorship of the Lick Observatory, whose splendid 36-inch refractor was then the most powerful telescope in the world, he proposed to observatory benefactor Darius O. Mills a hitherto untried method of searching for such a planet:

> Almost our only hope, . . . rests in . . . spectroscopic researches . . . since in the photographs that would be taken of all the stars, the spectrum of a planet would show instantly by its difference from the others. The Harvard College Observatory will probably not be able to secure such a discovery by this method because with a 12 in. telescope, which they are using, they cannot photograph the spectra of faint enough stars. With a 30, 33 or 36 [-inch] prism, we could probably photograph the spectra of all stars down to the 10th or 11th magnitude and I have very little doubt but that a planet exterior to Neptune, if it exists, would be brighter than the 10th magnitude. When I tell you that there are as many as 6,000,000 stars as bright or brighter than the 10th magnitude, you can see how hopeless the search for such a planet would be by any method other than the one which I have named.[1]

Holden's idea was actually hopelessly impractical at the time, and no spectroscopic searches for transneptunian planets were mounted at Lick or elsewhere. And yet the grand quest continued to attract sporadic attention from astronomers throughout the rest of the century. Predictions were made. Some, such as that by the Scottish astronomer George Forbes, were based on Camille Flammarion's 1879 suggestion that transneptunian planets might reveal their positions around points where the aphelia of periodic comets clustered. Another, by Copenhagen astronomer Hans-Emil Lau, followed a more classical line of attack based on the attempted analysis of the supposed residuals of Uranus (Neptune had not yet been under observation long enough to furnish helpful information). Both Forbes and Lau found indications of a pair of planets. These predictions, it must be admitted, were regarded as highly speculative even at the time. They inspired brief searches, but no planet was found.

The lack of success was hardly surprising. For one thing, there was nothing in the remaining residuals of Uranus remotely comparable to the gaping discrepancy of more than 2 arc minutes that led Le Verrier and Adams to their magnificent discovery. Celestial mechanicians could well reach different conclusions about the significance of any remaining residuals, and so they did. A well-publicized prediction was made in 1908 by Harvard's William Henry Pickering, although it was no more sophisticated than Todd's back-of-envelope calculation of 30 years earlier. Pickering simply did as Todd had done and graphed the residuals from Le Verrier's 1873 theory, and like him found a planet referred to as "Planet O" (to distinguish it from the several other planets Pickering later predicted). It was located at a distance of 51.9 AU from the Sun. However, it lay at a somewhat different heliocentric longitude from the planet Todd predicted.[2] The planet was searched for unsuccessfully at Harvard's Boyden Station in Peru and by amateur astronomer Joel Metcalf at Taunton, Massachusetts.

In 1909, Le Verrier's onetime associate, Gaillot, at the Paris Observatory, published his revised tables of the motions of Uranus and Neptune. It showed no clear evidence of perturbations by unknown planet or planets.[3] Later that year, however, Gaillot changed his mind and decided there might be two massive planets orbiting the Sun at distances of 44 and 66 AU.[4] Obviously, such vacillations did nothing to inspire public confidence.

Now one of the most energetic planet-seekers of all time entered the scene. Percival Lowell was an aristocratic Bostonian like Pickering. Harvard-educated, Lowell made a fortune in the family's textile mills by the time he was 30. He retired from business and spent most of the next 10 years in the Far East and then, enthused over the possibility Mars might be inhabited, founded in 1894 his own observatory in Flagstaff, Arizona. Possessed of incredible energy, he was soon in hot pursuit not only of sapient Martian life but also of a transneptunian planet. In contrast to his widely publicized and highly controversial Martian studies, he conducted the latter search "in virtual secrecy."[5] Lowell briefly tried the graphical methods of Todd and Pickering but became disillusioned with them. In the summer of 1910, he began a rigorous least-squares analysis of the residuals from Gaillot's theory of Uranus's motion. His approach was similar (given the smaller sizes of the residuals, which after Neptune had been included in the theory of Uranus never amounted to more than 4.5 arc seconds along Uranus's path) but necessarily much more tedious and complicated than Le Verrier had used in his analysis. Indeed it required massive effort not only from Lowell himself but from a team of human computers he hired over several years. His calculations went through several revisions. First he put the planet in Libra, but later his favored position moved into eastern Taurus—unfortunately in the heart of the Milky Way, with its myriad stars. With each revision, Lowell redirected his assistants to photograph the part of the sky where he had found the latest indications of a planet.

Though their searches failed to yield a planet, Lowell pressed on with his calculations, publishing them as his "Memoir on a Trans-Neptunian Planet" in 1915. The findings of his massive mathematical investigations were presented in carefully qualified words. Neptune, due to its relative nearness to Uranus and near circularity of its orbit had presented, according to Lowell, "the simplest possible case of the general problem."[6] That planet "turns out to have been most complaisant and to have assisted materially to its own detection." [7] Adams and Le Verrier proved to have been singularly fortunate, he declared. They had been able to use simplifying assumptions that Lowell could not allow himself. His "Planet X" probably followed a highly eccentric and inclined orbit, which put him and his computers on a head-on course with horrendous mathematical complications. More problematic was the fact that the residuals—the basis of all Lowell's deductions—could not be confidently relied on. Undoubtedly there remained uncertainties, perhaps considerable ones, in the theory of Uranus as well as systematic errors in the observations.

Despite the vagaries he encountered, Lowell finally offered two heliocentric longitudes for his Planet X, 180° apart. The most likely description of his planet was: mass about seven times that of the

Earth; mean distance from the Sun about 43 AU; orbit eccentric and inclined by perhaps 10° to the ecliptic. Hastened perhaps by over-work and discouragement—his Martian theories remained embat-tled, and his Planet X failed to turn up on his plates—Lowell died November 1916 of a massive brain hemorrhage.

Pickering, meanwhile, also remained active. He had pub-lished several more predictions for planets he called "P," "Q," and "R," their elements based on his statistical analysis of comet aphelia. P was located at a distance of 123 AU and had a period of 1,400 years; R, at a distance of 6,250 AU, had a period of half a million years and a mass 10,000 times that of the Earth; Q, at a dis-tance of 875 AU, had a period of 26,000 years and a mass of 20,000 earths, which made it "practically a dark companion to the Sun." These rather wild predictions did not attract serious comment from astronomers, although a brief unsuccessful search was mounted for his planet "O" by Milton Humason at Mt. Wilson Observatory in 1919.

Then there was a long hiatus. Nothing more was done until 1929, when astronomers at Lowell's observatory acquired a new wide-field photographic telescope more suitable for a planet search than the instruments that had been used by Lowell's assis-tants during his lifetime and a young astronomer, Clyde Tombaugh, to use to it. Since Lowell's death, his theoretical Planet X had moved from Taurus into Gemini. It was here that Tombaugh began to expose his plates. However, when Tombaugh found out how drastically Lowell's position had changed (Libra to Taurus), he began to regard the mathematical predictions as highly uncer-tain and of little real help.[8] Nevertheless, his thoroughness made up for his lack of faith. In February 1930, he discovered a planet, moving well beyond Neptune, which was later named Pluto. It was situated only 5.9° from the theoretical position calculated by Lowell for his Planet X (and, as Pickering hastened to point out, only 5.6° from where his Planet O was supposed to be lurking). Af-terwards it was realized that it had been registered on the Mt. Wil-son plates taken in 1919 during the brief search for Pickering's planet but went unrecognized; it had also shown up, but just barely, on Lowell Observatory plates from 1915.

The triumph of Adams and Le Verrier seemed, briefly, to have been repeated. It seemed Lowell had been vindicated posthumously; the icy world of his dreams really existed after all, and had turned up close to where his calculations had indicated it would be. Astronomers tended to discount Pickering's work, which had been much less rigorous. As a result, he spent the rest of his life a bitterly disappointed man.

The initial enthusiasm about the discovery was soon tempered. Was Pluto in fact Planet X? Later research proved not. Compared to the substantial gas giants sunwards of it, Pluto is a mere planetary soufflé—a creampuff planet, with a mass only 2/1000ths that of the Earth. Some astronomers doubt that it deserves to be regarded as a major planet at all. This is, admittedly, a minority view. In favor of retaining Pluto's planetary status it might be argued that Pluto does have a small moon—Charon, discovered only in 1978—and an atmosphere. In any case, it is certain it cannot be the object Lowell predicted from the residuals of Uranus's motion. The apparent fulfillment of Lowell's mathematical prediction was an illusion.

If so, then might Lowell's planet remain hidden among the stars, still awaiting discovery? Tombaugh himself renewed the search for additional planets after the discovery of Pluto. He covered 70 percent of the sky visible from the Lowell Observatory, including areas far from the ecliptic to make sure that planets with highly inclined orbits were not missed. He examined stars down to the 16th and 17th magnitudes. His search finally ended in 1943; he found no new planets and doubted that any were to be found. Given Tombaugh's thoroughness and skill, it was hard to imagine he could have missed anything. Moreover, a search by Charles T. Kowal using the 48-inch Schmidt telescope at Palomar, between 1977 and 1984, covered an area 15° on either side of the ecliptic. It failed equally to turn up any new planets, although it did reveal the unusual asteroid Chiron. Yet another search has been carried out by Thomas J. Chester and Michael Melnyk of the Jet Propulsion Laboratory, based on the observations of the Infrared Astronomical Satellite (IRAS) of 1983. The idea is that even a faint planet ought to stand out clearly in the infrared—thus almost reviving Holden's suggestion that unknown planets might be identified by their spectral characteristics. So far, as with other more orthodox searches, nothing has turned up.

This is hardly encouraging. And yet there are still the residuals of Uranus to be accounted for. If Pluto isn't massive enough to perturb the motion of Uranus, then must not another planet be pulling it off course? Not only that, but the early observations of Neptune were also apparently out of line; Galileo's prediscovery position from January 1613 was discrepant by 1 arc minute, while Michel Lalande's 1795 positions seemed to be out by at least 7 arc seconds.

Various astronomers have published calculations. U.S. Naval Observatory astronomers Thomas A. Van Flandern and Robert Harrington, working from an analysis of both Uranus and Neptune, proposed a hypothetical planet with a mass perhaps two to five times that of the Earth, orbiting between 50 and 100 AU from the Sun, with a period of several hundred years. Later Harrington revised his calculations and concluded for a planet with four times the mass of the Earth, located 2½ times Pluto's distance from the Sun. At the beginning of 1990, it was supposed to lie in Scorpius. Other planetary predictions were published by Conley Powell, an aerospace engineer in Alabama, whose planet was located in Virgo, and by Rodney S. Gomes and Sylvio Ferrz-Mello of Brazil, who found indications of several possible planets including one in a nearly circular orbit located at 45 AU from the Sun; it was located nearly 180° from Harrington's position in Cancer or Gemini (though with an uncertainty in its position of 50°).[9]

Clearly these predictions were more reminiscent of Pickering's than of Adams's and Le Verrier's. Probably, the residuals themselves are the problem. Using revised masses of Jupiter, Saturn, Uranus, and Neptune derived from the Voyager spacecraft missions and by readjusting the elements of the orbit of Uranus accordingly, E. Myles Standish of the Jet Propulsion Laboratory finds he is able to remove systematic trends in the residuals of that planet. Thus, according to Standish, there are no longer any significant anomalies in the motion of Uranus to be accounted for. "In hindsight," he concluded, "it seems apparent the residuals have not been correctly modeled in previous investigations. Either the orbit of Uranus was not adjusted at all, or it was adjusted incorrectly, or the mass of Neptune was determined incorrectly, or the whole question of Neptune's mass was ignored completely."[10] Standish has also expressed doubts about the accuracy of Galileo's and Lalande's positions.

Not everyone accepts Standish's conclusions—Van Flandern, for one, remains a holdout for systematic trends in the residuals.[11] On the whole, however, it seems probable that Planet X will go the way of Vulcan and enter finally into the realm of myth. And yet the work of the planet-seekers is far from over. The outer solar system contains the Kuiper belt of comets. The first objects, on the order of 1/10 the size of Pluto and lying outside its orbit, were identified by David Jewitt and Jane Luu in 1992.[12] Numerous other small bodies remain to be inventoried; when they are found, it will be by thorough searches similar to that by which Tombaugh found Pluto, not by the inverse-perturbation method of Adams and Le Verrier.

Then, too, there are other suns, other solar systems. The existence of planets beyond our system has recently been inferred from their perturbations in the motions of their parent stars 51 Pegasi, 70 Virginis, and 47 Ursae Majoris. Like Neptune, these planets have been "felt" before they have been seen. Indeed, they are much too faint to be recorded optically.

So the quest continues. The times are not yet fulfilled.

Chronology

U. J. J. LE VERRIER'S QUEST FOR VULCAN

1627	Kepler completes the *Rudolphine Tables*, which enables him to accurately predict the motions of Mercury and Venus, hence when they will transit or pass across the Sun.
1629	Kepler predicts a transit of Mercury for November 7, 1631.
1631	First recorded transit of Mercury observed by Gassendi.
1651	Riccioli comments adversely on the problem of Mercury.
1661	May 3. The third recorded transit of Mercury. Kepler's tables "neer the truth, and failed not many minutes."
1707	La Hire out by one day in his prediction of the transit of Mercury.
1753	La Hire and Edmond Halley err by several hours in their prediction of the transit of May 6.
1786	Lalande has a 53' error in his transit prediction.
1802	November 9. Transit widely observed; contact timings do not accord with theory.
1811	March 11. U. J. J. Le Verrier born Saint-Lô, Normandy.
1813	Lindenau publishes improved tables of Mercury. September 19. C. H. F. Peters born.

1814	August 11. Edmond Modeste Lescarbault born at Chateaudun.
1818	January 6. Capel Lofft watches an unknown dark object cross the Sun.
1820	February 29. Lewis Swift born at Clarkson, New York.
1826	Heinrich Schwabe commences solar observation in the hope of detecting a planet inside the orbit of Mercury.
1829	October 15. Asaph Hall born at Goshen, Connecticut.
1834	J. F. Benzenberg considers the possibility of intramercurial planets.
1835	Le Verrier graduates with distinction from École Polytechnique.
	March 12. Simon Newcomb born at Wallace, Nova Scotia.
1837	Lescarbault first attracted to the idea of unknown planets.
1838	January 28. J. C. Watson born.
	Thomas Dick publishes his speculation on the possibility of intramercurial planets.
1843	Le Verrier produces revised tables of Mercury. Unable to provide a suitable theory to account for the anomaly.
1845	May 8. Le Verrier's tables fail to accurately predict the transit of Mercury.
	Lescarbault's interest in unknown planets is rekindled by the sight of Mercury silhouetted on the face of the Sun.
	Benzenberg and J. F. J. Schmidt mount a telescopic search for interior planets.
	Le Verrier publishes his first memoir on the Uranus problem.
1846	J. Babinet suggests solar prominences are due to "incandescent clouds of a planetary kind, circling the Sun in the form of a train or portions of a ring." Proposes mythological name Vulcan. Analogous masses specified the Cyclopes.
	Dutch meteorologist C. H. D. Buys-Ballot attributes periodic variations in the mean temperature of the Earth's atmosphere to a ring of matter around the Sun inside the orbit of Mercury.
	Le Verrier publishes his second and third memoirs on Uranus.

September 18. Le Verrier solicits the help of Galle at Berlin to search for planet.

September 23. Galle and d'Arrest discover Neptune.

1847   Edward C. Herrick at Yale mounts a twice daily search of the Sun for a possible interior planet.

1848   Le Verrier's new tables fail to accurately predict the transit of Mercury.

1849   Le Verrier addresses the implications of the anomalous secular motion of Mercury. Holds back the incipient thought until he has reworked his data.

1850   Daniel Kirkwood indirectly promulgates the existence of intramercurial material through his Analogy, an empirical attempt to find a law governing the axial spin of the planets as Kepler's third law regulates their orbital motion.

1852   Benjamin Pierce hints at the possible existence of intramercurial planets.

Le Verrier publishes reductions of meridian-circle observations of the Sun taken at Greenwich.

1854   W. Thomson (Lord Kelvin) attributes the mechanical energies of the Sun to the infall of meteoric material. Le Verrier appointed director of the Paris Observatory and begins the monumental task of revising planetary theory.

1858   Le Verrier publishes revised tables of the Sun.

1859   Midafternoon March 26. Lescarbault observes transit of enigmatic dark spot across chord of the Sun.

September 12. Le Verrier publishes revised theory of Mercury. Uncovers a major problem. Missing mass hypothesis.

September. American weather prophet John H. Tice sees strange object in transit across the Sun.

Wolf publishes a list of suspect "sunspots."

December 22. Lescarbault writes to Le Verrier.

December 30. Le Verrier and a colleague descend unannounced on Lescarbault.

December 31. Le Verrier's informal account of the meeting with Lescarbault.

1860   January 2. Le Verrier presents formal account of discovery to the Academy of Sciences.

January 10. Benjamin Scott claims prior discovery in London *Times*.

January 15. Lescarbault awarded the Legion of Honor.

February 10. Radau refers to Lescarbault's object as Vulcan. Richard Carrington publishes list of suspect sunspots.

March. Liais attacks the observation. The Great Debate begins.

March and April. Expected transits of Vulcan do not materialize.

July 18. Total eclipse of the Sun; Vulcan not seen. Doubts mushroom.

Simon Newcomb has doubts reinforcing those already expressed by Liais.

1862　March 20. Dark spot seen on Sun by Lummis.

1865　Coumbary sees suspicious objects on Sun.

1869　Organized searches unsuccessful.

August 7. Nonstellar object suspected at the total solar eclipse.

1870　Charles Delaunay replaces Le Verrier, who has been obliged to resign his position, as director of Paris observatory.

1870–71　Franco-Prussian War. Paris besieged. Le Verrier quits capital.

1872　Autumn. Delaunay drowns in boating mishap. Le Verrier recalled to take his place. Resumes his immense labors in planetary theory.

1874　Le Verrier reaffirms his belief in Vulcan.

1876　April 4. Report of sighting from Weber in China. Interest revives.

Vulcan-mania grips readers of the *Scientific American*.

1877　Le Verrier issues his last alert.

September 23. Le Verrier dies in Paris.

Simon Newcomb becomes Superintendent of the Nautical Almanac Office, Washington D.C.

1878　July 29. The Great Solar Eclipse. Watson–Swift objects. C. H. F. Peters rejects the observations.

1879　Oppolzer predicts transit on March 18. Vulcan fails to appear.

1880   January 11. Rumor of a reddish object near the eclipsed Sun. November 22. Death of J. C. Watson.

1882   Tisserand dismisses single planet theory. Simon Newcomb examines the problem, finds the anomaly really does exist, but is 43″—larger than Le Verrier found.

1883   May 6. Caroline Island eclipse. Rumors of a very red star.

1884–85   Extensive visual monitoring of the Sun by a group of British amateurs.

1884   Julius Bauschinger rejects C. H. F. Peters's critique and reiterates Le Verrier's view.

1894   Asaph Hall proposes to modify the inverse-square law. August 11. Edmond Lescarbault dies at Orgères-en-Beauce.

1895   Newcomb reviews all the hypotheses advanced in explanation of the anomaly.

1900   After a 10-year photographic study of the Sun, E. C. Pickering announces no Vulcan brighter than magnitude 4 can exist.
        May. Suspect objects imaged at the total eclipse of the Sun.

1905   Albert Einstein introduces special theory of relativity.

1906   Hugo von Seeliger publishes his zodiacal light hypothesis. Widely accepted.
        First Trojan asteroid discovered.

1907   November 22. Death of Asaph Hall.

1909   Summing up the results of Lick Observatory solar eclipse searches of 1901, 1905, and 1908, W. W. Campbell announces the closing of the problem. Favors Seeliger's idea.
        July 11. Death of Simon Newcomb.

1912   Trojan Vulcans advocated by C. V. L. Charlier.

1913   January 5. Lewis Swift dies in Cortland, near Syracuse.

1914   Charlier fails in his attempt to photograph the libration points of Mercury.

1915   Einstein completes general theory of relativity and accounts for the anomalous advance of Mercury's perihelion. The Newtonian or missing mass solution now unnecessary.

1921   W. M. Smart investigates the possible existence of Trojan Vulcans.

1923   Robert Trumpler follows up Charlier's suggestion. Campbell and Trumpler sweep suspect zone because of the interest inherent in finding new bodies. None found.

1929   Photographs of the May total eclipse of the Sun taken in Sumatra by E. Freundlich examined by H. von Klüber. No suspects found. The search for Vulcan finally over.

Notes and References

Introduction

1. Watson Warren Zachary, "*An Historical Analysis of the Theoretical Solutions to the Problem of the Advance of the Perihelion of Mercury,*" Ph.D. thesis (University of Wisconsin: 1969, unpublished), 18.
2. N. R. Hanson, "Leverrier: The Zenith and Nadir of Newtonian Mechanics," *Isis* **53,** 359 (1962).
3. Y. Villarçeau, "Discours au nom des Astronomes de l'Observatoire," *Compte Rendu des Séances de l'Académie des Sciences* **85,** 584 (1877).
4. U. J. J. Le Verrier, "Recherches sur les mouvements d'Uranus," *Compte Rendu des Séances de l'Académie des Sciences* **22,** 907–918 (1846); henceforth given as *Compte Rendu*
5. N. R. Hanson, Ref. 2, 359.
6. U. J. J. Le Verrier, "Détermination nouvelle de l'orbite de Mercure et de ses perturbations," *Compte Rendu* **16,** 1054 (1843).
7. David Brewster, "Recent Discoveries in Astronomy," *North British Review* **33,** 6 (1860).

Chapter 1

1. E. M. Antoniadi, *The Planet Mercury;* trans. Patrick Moore (Devon: Keith Reid, 1974), pp. 9–11. For the reference to Mercury the elusive, Antoniadi cites the authority of Proclus Diadochus; for Mercury the nimble one, Apuleius.
2. Owen Gingerich, "Ptolemy and the Maverick Motion of Mercury," *Sky and Telescope* **66,** 12 (1983).
3. Copernicus, "On the Revolutions of the Heavenly Spheres," Book V, sec. 29.
4. This is conventionally measured from a fixed point in space known as the first point of Aries, γ, where the ecliptic cuts the equator going from south to north.

5. The ascending node is the point where the planet crosses the ecliptic traveling north; the descending node is where the planet crosses the ecliptic traveling south.

6. To give an example, for Mars Kepler derived the following elements: the semi-major axis a = 1.5264 AU; eccentricity, e = 0.0926; longitude of perihelion, inclination, i = 1° 50′ 25″; longitude of perihelion, $\tilde{\omega}$ = 328° 48′ 55″; and longitude of ascending node, Ω = 46° 46½′.

7. For the significance of the observation see Albert Van Helden, "The Importance of the Transit of Mercury of 1631," *Journal for the History of Astronomy* **7**, 1–10 (1976), and Robert Grant, *History of Physical Astronomy from the Earliest Ages to the Middle of the Nineteenth Century* (London: Henry G. Bohn, 1852), pp. 415–417.

8. P. Humbert, *L'Oeuvre astronomique de Gassendi* (Paris, 1936), 23.

Chapter 2

1. For information on Horrocks, the following have been consulted: Rev. Arundell Blount Whatton, *Memoir of the Life and Labours of the Rev. Jeremiah Horrox* (London, 1859), J. E. Bailey, *The Palatine Note-books*, abridged in *The Observatory* **6**, 318–328 (1883), Allan Chapman, *Three North Country Astronomers* (Manchester: Neil Richardson, 1982), and Chapman, "Jeremiah Horrocks, the transit of Venus, and the 'New Astronomy' in Seventeenth Century England," *Quarterly Journal of the Royal Astronomical Society* **31**, 333–357 (1990).

2. Robert Grant, *History of Physical Astronomy* (London: Henry G. Bohn, 1852) p. 424.

3. Chapman, *Three North Country Astronomers*, 36.

4. Vincent Wing, *An Ephemerides of the Coelestial Motions for XIII Years, beginning Anno 1659, ending Anno 1671* (London, 1658).

5. William Stukeley, *Memoirs of Newton's Life . . . 1752; Being Some Account of His Family and Chiefly of the Junior Part of His Life* (London: Taylor and Francis, 1936), pp. 46–47.

6. Quoted in Richard Westfall, *The Life of Isaac Newton* (Cambridge: Cambridge University Press, 1993), 22.

7. Ibid., 23.

8. A. Rupert Hall, "Newton's Note-book, 1661–1665," *The Cambridge Historical Journal* **IX** (no. 2), 249 (1948).

9. D. T. Whiteside, ed., *The Mathematical Papers of Isaac Newton* (Cambridge: Cambridge University Press, 8 volumes, 1967–1980), vol. 1, p. 148.

10. Stukeley, ibid., 82.

11. Westfall, *Never at Rest*, p. 51. The other accounts are given by Henry Pemberton, *A View of Sir Isaac Newton's Philosophy* (London, 1728), preface and a MS by Newton himself, which reads: "I began to think of gravity extending to ye orb of the Moon & . . . from Keplers rule of the periodical times of the Planets being in sesquilaterate proportion of their distances from the center of their Orbs, I deduced that the forces wch keep the Planets in their Orbs must [be] reciprocally as the squares of their distances from the centers about wch they revolve: & thereby compared the force requisite to keep the Moon in her Orb with the force of gravity at the surface of the earth, & found them answer pretty nearly."

12. Hooke to Newton, January 17, 1680; *Correspondence*, vol. 2, p. 313.

13. *Principia*, Book I, proposition i, theorem i; the converse, given the areal law, that the force must be directed toward the center, is proved in the next proposition.
14. *Principia*, Book I, propositions x and xi.
15. Westfall, *Never at Rest*, p. 403.
16. Quoted by Wilson in "The Newtonian Achievement in Astronomy," in R. Taton and C. Wilson, editors, *Planetary Astronomy from the Renaissance to the Rise of Astrophysics, Part A: Tycho Brahe to Newton* (Cambridge: Cambridge University Press, 1989), 253.
17. Quoted in Westfall, *Never at Rest*, 406.
18. Quoted in Ivars Peterson, *Newton's Clock: Chaos in the Solar System* (New York, W. H. Freeman and Company, 1993), 87.
19. Letter of Newton to Bentley, Jan. 17, 1692–93; in *Works of Richard Bentley* (London, 1838), vol. 3, pp. 210–211.
20. Ibid.
21. Grant, *History*, 30.
22. Ibid., 38.
23. *Principia*, Book I, prop. XLI, problem xxi.
24. Sir John F. W. Herschel, *Outlines of Astronomy* (New York, 10th ed., 1857), 411.
25. Westfall, *Never at Rest*, 443.
26. Seth Ward, Savilian Professor of Astronomy at Oxford; quoted in Frank E. Manuel, *A Portrait of Isaac Newton* (Cambridge: Belknap Press of Harvard University Press, 1968), 294.
27. David Gregory, memorandum September 1, 1694, *Correspondence*, IV, 7.
28. Newton to Flamsteed, April 23, 1695; *Correspondence*, IV, 106.
29. Newton to Flamsteed, July 9, 1695; *Correspondence*, IV, 143.
30. Whiteside, "Newton's Lunar Theory: From High Hope to Disenchatment," *Vistas in Astronomy*, 19 317-328:320 (1976).
31. Ibid., p. 324.

Chapter 3

1. Voltaire, Correspondence, April 2, 1764; quoted in Will Durant, *The Story of Philosophy* (New York: Washington Square Press, 1961), p. 246.
2. A. Pannekoek, *A History of Astronomy* (London: George Allen & Unwin Ltd., 1961; New York: Dover, 1989 reprint), p. 298.
3. Pannekoek, 299.
4. Pannekoek, 300.
5. Quoted in Ivars Peterson, *Newton's Clock: Chaos in the Solar System* (New York: W. H. Freeman and Company, 1993), pp. 133–134.
6. Ibid., 135.
7. Another second-order effect, independent of the Sun's motion, is also important. As described by Airy, *Gravitation*, p. 80: "When the line of apsides is directed toward the Sun, the whole effect of the force is to make it progress, that is, to move in the same direction as the Sun: the Sun passes through about 27° in one revolution of the Moon, and therefore departs only 16° from the line of apsides; and therefore the apsides continues a long time near the Sun. When at right angles to the line joining the Earth and Sun, the whole effect of the force is to make it regress, and therefore, moving in the direction opposite to the

Sun's motion, the angle between the Sun and the line of apsides is increased by 36° in each revolution, and the line of apsides soon escapes from this position. The effect of the former is therefore increased, while that of the latter is diminished." As in the other case, a small addition to the apogeal over the perigeal effect produces a substantial increase in the effective motion.

8. Grant, *History of Physical Astronomy*, 46.

9. E. Halley, *Astronomical Tables with Precepts both in English and Latin* (London, 1752).

10. See Peter Broughton, "The First Predicted Return of Comet Halley," *Journal for the History of Astronomy* **XVI**, 123–133 (1985), and Curtis Wilson, "Clairaut's Calculation of the Eighteenth-Century Return of Halley's Comet," *Journal for the History of Astronomy* **XXIV**, 1–15 (1993).

11. J. J. le F. de Lalande, *Bibliographie astronomique avec l'histoire de l'astronomie depuis 1781 jusqu'à 1802* (Paris, 1803), 677.

12. Clairaut, *Théorie du mouvement des comètes, dans laquelle on a égard aux altérations que leurs orbites éprouvent par l'action des planètes. Avec l'application de cette théorie à la comète qui a été observée dans les années 1531, 1607, 1682, & 1759* (Paris, 1760), 5.

13. Thomas Carlyle, "Pen Portraits," in *Carlyle: Representative Selections*, ed. A. W. Evans (London: G. Bell & Sons, 1913), pp. 363–364.

Chapter 4

1. Quoted in E. T. Bell., *Men of Mathematics* (New York: Simon and Schuster, 1937), p. 153.

2. W. W. Rouse Ball., *A Short Account of the History of Mathematics* (New York: Dover, 1960; reprint of 4th ed., 1908), p. 411.

3. Bell, p. 173.

4. The term secular is, incidentally, taken from the Latin *saeculaires*, occurring once in or lasting an age; ancient Roman games or festival held at long intervals.

5. Laplace, *Mém. Acad. des Sciences*, 1784.

6. Laplace, *System of the World*, Vol. II, p. 32.

7. Lagrange, *Mém. Acad. des Sciences de Berlin*, 1776.

8. Taton and Wilson, eds., *Planetary Astronomy from the Renaissance to the Rise of Astrophysics, Part B: the Eighteenth and Nineteenth Centuries* (Cambridge, Cambridge University Press, 1995), 46.

9. Grant, *History of Physical Astronomy*, 56.

10. Thomas Carlyle, *Sartor Resartus: The Life and Opinions of Herr Teufelsdröckh*, C. F. Harrold (New York: The Odyssey Press, 1939), pp. 257–258.

11. Grant, *History of Physical Astronomy*, 56.

12. Ibid.

13. Laplace, *System of the World*, Vol. II, p. 35.

Chapter 5

1. An interesting account is by Joseph Ashbrook, "An alleged satellite of Venus," *The Astronomical Scrapbook: Skywatchers, Pioneers, and Seekers in Astronomy* (Cambridge, Massachusetts: Sky Publishing, 1984), p. 281.

2. Alexis Claude Clairaut, "Memoire sur la comete de 1682," *Journal des sçavans*, **40**, 38–45 (Jan. 1759).

3. William Herschel, "Account of a Comet," *Phil. Trans.* **lxxi**, 492–501 (1781); in J. L. E. Dreyer, *Scientific Papers of William Herschel* (London, Royal Astronomical Society, 1912), vol. 1, p. 30. A useful account of the discovery and its aftermath is Simon Schaffer, "Uranus and the Establishment of Herschel's Astronomy," *Journal for the History of Astronomy* **XII**, 11–26 (1981). See also J. A. Bennett, E. G. Forbes, M. Hoskin, R. Porter, and R. W. Smith, "History of the Discovery of Uranus," in *Uranus and the Outer Planets*, ed. Garry Hunt (Cambridge: Cambridge University Press, 1982), pp. 21–89.

4. Herschel's set of measures, with the magnifications used on each occasion, are discussed in an interesting paper by R. H. Austin, "Uranus Observed," *The British Journal for the History of Science* **iii**, 275–284 (1967). Austin points out Herschel's measures suggested the diameter was increasing and that the comet was approaching, at a time when as we now know, the apparent diameter was actually decreasing, and concludes that Herschel's measures "recorded the readings as a trend and the cause of the trend was Herschel's expectation." Herschel's preconceptions about the object's cometary nature no doubt played a role, but so, less profoundly, did "several optical deceptions" which Herschel himself noted, e.g., the fact that "a very small object will appear something less in a telescope when we see it first than when we become familiar with it."

5. Constance A. Lubbock, ed., *The Herschel Chronicle* (Cambridge: Cambridge University Press, 1933), p. 80.

6. Ibid., 86.

7. Ibid., 94.

8. A. F. O'D. Alexander, *The Planet Uranus* (New York: American Elsevier, 1965), pp. 31, 51.

9. Lubbock, 93.

10. Ibid., 95.

11. Ibid.

12. Herschel, *Scientific Papers*, 100–101.

13. Winston Churchill, *History of the English-Speaking Peoples*, (London: Cassell, 1995 reprint), vol. 3, p. 142.

14. Matthew Turner, in *The Theological and Miscellaneous Works of Joseph Priestley*, ed. J. T. Rutt (London, 1817) pp. i, 76 as quoted in Schaffer, "Uranus and the Establishment of Herschel's Astronomy," *Journal for the History of Astronomy*, **XII**, 11–26: 23 (1981).

15. Ibid., 15.

16. Lubbock, 76.

17. Schaffer, 15.

18. A single observation of Uranus by Flamsteed, or possibly his assistant Crosthwaite, in 1714 was found as recently as 1968 by Dennis Rawlins. See "A long lost observation of Uranus, Flamsteed 1714," *Publications of the Astronomical Society of the Pacific* **80**, 217–219 (1968).

19. Alexander, pp. 87–89. Two more observations of Uranus by Bradley, made in 1748 and 1750, were discovered in 1864.

20. Dennis Rawlins, "The Unslandering of Sloppy Pierre," *Astronomy* **9**, 24–28 (1981). Above an observation he made December 27, 1768, Lemonnier wrote clearly in

minute handwriting, "c'est la nouvelle planète découverte sur 1781 la 13 Mars par Herschell" (It is the new planet discovered on March 13 by Herschel).

21. For the history of Bode's law, see Michael Martin Nieto, *The Titius-Bode Law of Planetary Distances* (Oxford: Oxford University Press, 1972); Stanley L. Jaki, "The Early History of the Titius-Bode Law," *American Journal of Physics* **40,** 1014–1023 (1972), "The Original Formulation of the Titius-Bode Law," *Journal for the History of Astronomy* **iii,** 136–138 (1972), "The Titius-Bode Law: A Strange Bicentary," *Sky and Telescope* **43,** 280–281 (1972). Though Bode cited the relationship in the next two editions of his primer and in a 1778 publication, he failed to credit Titius until 1784—after the discovery of Uranus.

22. See Clifford J. Cunningham, "The Baron and his Celestial Police," *Sky and Telescope* **75,** 271–272 (1988).

23. The account of the 1800 meeting at Lilienthal is based on the following publications by Dieter Gerdes of the Schroeter Museum, Lilienthal: *Die Lilienthaler Sternwarte 1781 bis 1818: Machinae Coelestes Lilienthalienses die Instrumente, eine zeigeschicthliche Dokumentation* (Lilienthal: Heimatrerein, 1991) and *Die Geschichte der Astronomischen Gesellschaft gegrundet in Lilienthal am 20 September 1800. Die ersten 63 Jahre ihres Bestehens von 1800 bis 1863* (Lilienthal: Heimatrerein, 1990). For a brief popular account of Schroeter's life and work, see William Sheehan and Richard Baum, "Observation and Inference: Johann Hieronymus Schroeter, 1745–1816," *Journal of the British Astronomical Association* **105,** 171–175 (1995). The events of April 20, 1813, are recounted by Richard Baum in "The Lilienthal Tragedy," *Journal of the British Astronomical Association* **101,** 369–370 (1991).

24. Johann Elert Bode, *Von dem neuen, zwichen Mars und Jupiter entdeckten Hauptplaneten des Sonnensystems* (Berlin, 1802).

25. Franz Xaver von Zach, "Über einen zwischen Mars and Jupiter längst vermutheten, nun wahrscheinlich entdeckten neuen Hauptplaneten unseres Sonnen-Systems," *Monatliche Correspondenz zur Beförderung der Erd und Himmelskunde* **3,** 592–623 (1801).

26. Eric G. Forbes, "Gauss and the Discovery of Ceres," *Journal for the History of Astronomy* **ii,** 195–199 (1971).

27. Lubbock, p. 273.

28. G. B. Airy, "Address Delivered by the Astronomer Royal, President of the Society, on Presenting the Honorary Medal of the Society to Dr. Annibale de Gasparis," *Monthly Notices of the Royal Astronomical Society* **11,** 119 (1851).

29. H. Goldschmidt, obituary notice, *Monthly Notices of the Royal Astronomical Society* **27,** 115–117 (1867).

Chapter 6

1. Bell, 181.

2. Ibid.

3. Ibid., 176.

4. Ibid., 181. To which Napoleon replied: "Ah, but that is a fine hypothesis. *It explains so many things.*"

5. Zachary, p. 12.

6. Laplace, *The System of the World*, Vol. II, p. 54.

7. As suggested by Jacques Merleau-Ponty in "Laplace as Cosmologist," in *Cosmology, History, and Theology,* ed. W. Yougrau and Allen D. Breck (New York: Plenum Press, 1977), pp. 282–291, referring to the influential scheme of scientific progress developed by Thomas Kuhn in *The Structure of Scientific Revolutions* (Chicago: University of Chicago Press, 1961).

8. Bruno Morando, "Laplace," in *Planetary Astronomy,* ed. René Taton and Curtis Wilson, 150.

9. Laplace, *System,* Vol. II, p. 20.

10. Ibid., 2.

11. Zachary, 56.

12. Robert W. Smith, "The Cambridge Network in Action: The Discovery of Neptune," *Isis* **80,** 395–422:398 (1989).

13. Morton Grosser, *The Discovery of Neptune* (Cambridge: Harvard University Press, 1962), p. 58.

14. Joseph Bertrand, "Élogie historique de Urbain-Jean-Joseph Leverrier," *Annales de l'Observatoire de Paris, Mémoires* **15,** 3–22:5 (Paris, 1880)

15. Ibid., 6.

16. Ibid.

17. U. J. J. Le Verrier, "Sur les variations seculaires des orbites des planetes," *Comptes Rendus* **9,** 370–374 (1839).

18. E. Dunkin, "M. Le Verrier," *The Observatory* **7,** 199–206:201 (1877).

19. Zachary, 7–8.

20. Ibid., 16.

21. O. M. Mitchel, *Orbs of Heaven* (London: G. Routledge & Co., 1857), pp. 138–139.

22. U. J. J. Le Verrier, "Sur la comète observée de M. Faye, 1843, Nov. 22 et sur son identité avec la comète de Lexelle," *Comptes Rendu* **18,** 826–827 (1844), and "Calcul de la valeur des perturbations que la comète découverte par De Vico, 1844, Aug. 22, peut éprouver par l'action de la Terre," *Comptes Rendu* **19,** 666–670 (1845).

Chapter 7

1. Smith, "Cambridge Network," 398.

2. This tantalizing remark appears in George Chambers, *A Handbook of Descriptive and Practical Astronomy* (Oxford, 4th ed., 1889), vol. 1, p. 253 footnote. This statement apparently depends upon a note found among Lalande's papers presented to the Académie des Sciences in 1852.

3. Rawlins, "The Unslandering of Sloppy Pierre."

4. Alexis Bouvard, *Tables astronomiques publiées par le Bureau des Longitudes de France contenant les Tables de Jupiter, de Saturne et d'Uranus construites d'après la théorie de la Mécanique céleste* (Paris, 1821), p. xiv.

5. Grosser, p. 42. The literature on the discovery of Neptune is large. Grosser's account is still essential and has been largely followed by later writers, not always with due acknowledgement. Other works that deal extensively with the discovery of Neptune include Robert Grant, *History of Physical Astronomy* (London: Henry G. Bohn, 1852); H. Spencer Jones, *John Couch Adams and the Discovery of Neptune* (Cambridge: Cambridge University Press, 1947); W. M. Smart, "John Couch Adams and the Discovery of Neptune," *Nature* **158,** 648ff (1946); also the

Reply by Spencer Jones, 158, p. 830; W. M. Smart, "John Couch Adams and the Discovery of Neptune," *Occasional Notes, Royal Astronomical Society* no. 2; 33ff (1947); H. Spencer Jones, "G. B. Airy and the Discovery of Neptune," *Popular Astronomy* **55,** 312–316 (1947) (which includes a reply by W. M. Smart); W. G. Hoyt, *Planets X and Pluto* (Tucson: University of Arizona Press, 1980); Allan Chapman, "Private Research and Public Duty: George Biddell Airy and the Search for Neptune," *Journal for the History of Astronomy* **xix,** 121–139 (1988); Mark Littman, *Planets Beyond: Discovering the Outer Solar System* (New York: Wiley, 1988); Patrick Moore, *The Planet Neptune* (New York: Wiley, 1988); Robert W. Smith, "The Cambridge Network in Action: The Discovery of Neptune," *Isis* **80,** 395–422 (1989); J. G. Hubbell and Robert W. Smith, "Neptune in America: Negotiating a Discovery," *Journal for the History of Astronomy* **xxiii,** 261–291 (1992).

6. Smith, "Cambridge Network," 399: "The possible amendment he examined involved specific attraction, that is, the notion that there might be something specific to or dependent on the nature or chemical constitution of a body that affects its gravitational attraction."

7. Bessel, *Populäre Vorlesungen,* 448; quoted in Grosser, 44.

8. George Biddell Airy, "Report on the Progress of Astronomy during the Present Century," *Report on the First and Second Meetings of the British Association for the Advancement of Science* (London, 1833), 189.

9. Such a medium was being invoked about this time by Johann Franz Encke to account for the fact that the comet now named for him (P/Encke, with the shortest period of any comet known, 3.3 years) showed a slight deceleration with each revolution, even after perturbations of the planets had been allowed for. Between 1789 and 1838, the period of Encke's comet decreased by 1.9 days, which came out to 0.1 day per revolution. The fact that other comets, such as P/Halley, were delayed in their returns was inconsistent with the resisting medium idea.

10. Smith, "Cambridge Network," 398.

11. G. B. Airy, "Account of Some Circumstances Historically Connected with the Discovery of the Planet Exterior to Uranus," *Monthly Notices of the Royal Astronomical Society* **7,** 123 (1846).

12. Ibid., 124.

13. Grosser, 50–51.

14. W. H. Smyth, "Extract of a letter from Capt. Smyth to the President, Containing the Translation of a Notice from M. Cacciatore," *Monthly Notices of the Royal Astronomical Society* **3,** 139 (1835). See also N. Cacciatore, "Sur une nouvelle petite planète dont l'existence a été soupçonnée par M. Cacciatore, directeur de l'Observatoire de Palerme," *Comptes Rendu* **2,** 154–155 (1836). See also R. Baum, *The Planets: Some Myths and Realities* (Newton Abbot: David & Charles, 1973), 163–168.

15. For Wartmann's report, see "Lettre de M. Wartmann, de Geneve, à M. Arago, sur un astre ayant l'aspect d'une étoile et qui cependant était doue d'un mouvement propre," *Comptes Rendu* **2,** 307–311 (1836). Hind's critique was published as "Wartmann's Supposed Planet," *Monthly Notices of the Royal Astronomical Society* **7,** 274 (1847). See also Baum, *The Planets,* Chap. 8. Theodore Oppolzer finally explained the Wartmann affair in 1880, pointing out that on average the positions given by Wartmann were 6 minutes of time greater in

right ascension and about 26 arc seconds farther north than the corresponding places of Uranus. "Whether Wartmann erred in plotting a map or in reading positions from a map, or whether precession was incorrectly applied is difficult to decide. Nevertheless, it is remarkable that subtraction of the approximate precession in a century removes the discrepancy." See *Astronomische Nachrichten* **97,** 253 (1880). Meanwhile Cacciatore's object remains unexplained.

16. Airy, "Account," 125.

17. Ibid.

18. B. A. Gould, Jr., *Report of the History of the Discovery of Neptune* (Washington, DC: Smithsonian Institution, 1850), p. 1.

19. F. W. Bessel, "Über die Verbindung der astronomischen Beobachtungen mit der Astronomie," in *Populäre Vorlesungen uber wissenschaftliche Gegenstäde* (Hamburg, 1848), pp. 408–457.

20. Smith, "Cambridge Network," 399.

21. Bouvard's *Tables* were never published. The details cited here are given in the account of his colleague Emmanuel Liais, *L'espace celeste et la natur tropicale* (Paris: Garnier Frères, 1866). The authors are indebted to Patrick Moore for making available a copy of his translation of part of this work, "History of the Discovery of the Planet Neptune."

22. *Comptes Rendu* **21,** 1050–1055 (1845).

23. Le Verrier pointed out that Bouvard's methods for computing the eccentricity of the orbit of Uranus were inconsistent, and yielded three widely different answers, that his first two tables disagreed in the rate of the secular motion of the mean longitude, and that the work contained an unforgivable number of typographical errors.

24. Le Verrier, abstract of "Recherches sur les mouvements d'Uranus,"*Comptes Rendu* **22,** 907–908 (1846).

25. He had technical reasons for approaching the problem in this manner. As explained by Lyttleton, even with the distance a/a' settled, "rightly or wrongly, there still remain *eight* unknowns, and quite apart from the large task of solving equations of conditions, not all these remaining eight quantities occur linearly. The principal difficulty arises from the unknown ϵ' [the heliocentric longitude]." To meet this difficulty, "Le Verrier . . . took 40 values of ϵ' at 9° intervals, so that the whole possible range of from 0° to 360° was covered. For each assumed value of ϵ', the expression for P(t) immediately becomes linear . . . and it is then possible to solve the equations of condition by least-squares. . . . This was Le Verrier's method, but clearly much arithmetical labour is involved in such an approach to the problem." R. A. Lyttleton, "The rediscovery of Neptune," *Vistas in Astronomy* **3,** 27–28 (1960).

26. Gould, 32–33.

27. Hoyt, 49.

28. For biographical details concerning Adams, we have followed J. W. L. Glaisher, "Biographical Notice," in J. C. Adams, *The Scientific Papers of John Couch Adams,* ed. W. G. Adams (Cambridge: Cambridge University Press, 1896), vol. 1, pp. xv– xlviii; Morton Grosser, *The Discovery of Neptune* (Cambridge, Harvard University Press, 19xx), xx; and H. M. Harrison, *Voyager in Time and Space: The Life of John Couch Adams, Cambridge Astronomer* (Lewes, Sussex: The Book Guild, 1994), xx.

29. Quoted in Grosser, 72.

30. Written by a fellow undergraduate at St. John's, William Wordsworth, in his *Prelude*, III, 60–64. Wordsworth matriculated in 1791.

31. Harrison, 73.

32. J. C. Adams, "An Explanation of the Observed Irregularities in the Motion of Uranus, on the Hypothesis of Disturbances Caused by a More Distant Planet; with a Determination of the Mass, Orbit, and Position of the Disturbing Body," *Memoirs of the Royal Astronomical Society* **16** (1847); in J. C. Adams, *Scientific Papers*, vol. 1, p. 8.

33. Harrison, 19–20.

34. J. C. Adams, "Letter on de Vico's Comet," *Times* (London), October 15, 1844. This comet, with a period of somewhat more than 5 years, was discovered by de Vico at Rome on August 23, 1844. Subsequently it was lost, but recovered by Edward Swift in 1894. Hence, it is known as Comet de Vico-Swift. It is not to be confused with another comet discovered by de Vico on February 20, 1846, with period 75 years. Although missed at its expected return in 1922, it was recovered in 1995.

35. J. C. Adams to George Adams, July 10, 1845; quoted in Grosser, 86.

36. James Challis, "Account of Observations at the Cambridge Observatory for Detecting the Planet Exterior to Uranus," *Monthly Notices of the Royal Astronomical Society* **7,** 121–149:148 (1846).

37. Airy, "Account," 129.

38. As noted, for instance, in the account by his brother Thomas, "he took with him [his calculations] on his way to Cambridge to the Astronomer Royal at Greenwich. . . . The result is well known. They were neglected. John was terribly disappointed and annoyed, for which he had great reason." Harrison, *Voyager,* 20.

39. E. W. Maunder, in *The Royal Observatory Greenwich* (London, 1900), 116.

40. Ibid., 118.

41. Allan Chapman, "Private Research and Public Duty: George Biddell Airy and the Search for Neptune," *Journal for the History of Astronomy* **xix,** 134 (1988).

42. Ibid., 126.

43. G. B. Airy to J. C. Adams, November 5, 1845; see Airy, "Account," 130.

44. Chapman, "Airy and Neptune," 127–128.

45. Ibid., 128. Even then he did not say he would have actually *looked* for the planet.

46. The query was not, in fact, as trivial as Adams had regarded it, as noted, for instance, in J. E. Littlewood, *A Mathematician's Miscellany* (London, 1953), 131. Adams himself eventually was forced to admit he had perhaps "hastily inferred" that the hypothesis of an exterior planet would *automatically* satisfy the errors in the radius vector.

47. *Comptes Rendu* **22,** 907–918 (1846).

48. Airy, "Account," p. 132.

49. Quoted in Smith, "Cambridge Network," 404.

50. Airy, "Account," 133.

51. Le Verrier to Airy, June 28, 1846; in Airy, "Account," 134.

52. Airy, "Account," 135.

53. After the discovery of Neptune, he explained, rather awkwardly, to Le Verrier: "I do not know whether you are aware that collateral researches had been go-

ing on in England and that they led to precisely the same results as yours." When asked to defend his actions (or inactions) by his friend Adam Sedgwick at Cambridge, he wrote, "When about June last Le Verrier published one of the results Adams had attained before (September 1845), why in the name of wonder was not all Europe made to ring with the fact that a B.A. at Cambridge had done this 10 [sic] months previously?

"In the name of wonder what had *I* to do with this publication. No understood rule of Society would have justified me in doing so. The *first* person to publish was Adams. The *second* was Challis. The third was I. But there was a very serious difficulty in the way of *my* doing so, because Adams had declined to answer my letter. Moreover, in consequence of my question not having been resolved, I had not till I received Le Verrier's explanatory letter the security for the truth of the theory which I desired, . . ." Quoted in Harrison, *Voyager*, 69.

54. Ibid., 136.
55. Airy, "Account," 136.
56. Ibid., 137.
57. Airy, "Account," 148.
58. Le Verrier, "Sur la planète qui produit les anomalies observées dans le mouvement d'Uranus-détermination de sa masse, de son orbite, et le sa position actuelle," *Comptes Rendu* **23**, 428–438:433 (Aug. 31, 1846).
59. J. F. W. Herschel, "Le Verrier's Planet," letter, *Athenaeum* **1019** (Oct. 3, 1846).
60. Smith, "Cambridge Network," 411.
61. Ibid., 409, Rev. Richard Sheepshanks, Nov. 12, 1846.
62. Ibid., 408, J. R. Hind to James Challis, September 16, 1846.
63. William Rowan Hamilton recounted Herschel's comments to Dawes to a correspondent, October, 1846; see Robert Perceval Graves, *Life of William Rowan Hamilton*, (Dublin/London, 1885), vol. II, p. 529.
64. E. S. Holden to Jane Lassell, May 9, 1890; Mary Lea Shane Archives of the Lick Observatory. The first published mention of Lassell's missed opportunity was by E. S. Holden, "Historical Note Relating to the Search for the Planet Neptune in England, 1845–46," *Astronomy and Astro-Physics* **11**, 287 (1892). Holden however assumed the date of Dawes' letter as September 1845. Robert W. Smith considered Lassell's participation in "William Lassell and the Discovery of Neptune," *Journal for the History of Astronomy xiv*, 3 (1983), but misses an interesting note by A. Marth, "Report of the Meeting of the British Astronomical Association," *Journal of the British Astronomical Association* **2**, 433–434 (1892). Marth's source was also Mrs. Lassell. He understood the date of Dawes's letter was September 1846—only a couple of weeks before the discovery of Neptune—which nicely squares the episode with other developments. The whole Lassell episode is reconsidered by Richard Baum, "William Lassell and 'the Accident of a Maid-servant's Carelessness' or Why Neptune was not Searched for at Starfield," *Journal of the British Astronomical Association* **106**, 217–219 (1996).
65. H. H. Turner, obituary notice of Johann Gottfried Galle, *Monthly Notices of the Royal Astronomical Society* **71**, 275–281:278 (1911).
66. J. L. E. Dreyer, "Historical Note Concerning the Discovery of Neptune," *Copernicus* **2**, 63–64 (1882).

Chapter 8

1. Edwin Holmes, "The Planet Neptune," *Journal of the British Astronomical Association* **18**, 36–37 (1907).
2. This was slightly too large; the actual value on the date in question, according to modern measures, would have been 2".5. This corresponds to a diameter of 49,530 kilometers, compared with 12,756 kilometers for the Earth.
3. H. H. Turner, obituary notice of Galle, 280.
4. Grosser, 124.
5. J. Encke, "Letter to the Editor," *Astronomische Nachrichten* **580**, (1846). For other accounts of the discovery, see Encke, "Account of the Discovery of the Planet of Le Verrier at Berlin," *Monthly Notices of the Royal Astronomical Society* **7**, 153 (1846), and J. Galle, "Ueber die erste Auffindung des Planeten Neptun," *Copernicus* **2**, 96 (1882).
6. J. Challis, "The Search for the Planet Neptune by Professor Challis," *Astronomische Nachrichten* **26**, 101–106 (1846).
7. F. Arago, "Letter about the Name of Neptune," *Astronomische Nachrichten* **25**, 81 (1846).
8. Arago, *Astronomische Nachrichten* **25**, 159 (1847).
9. Hoyt, *Planets X and Pluto*, 53.
10. H. Spencer Jones, *John Couch Adams and the Discovery of Neptune* (Cambridge: Cambridge University Press, 1946), 39.
11. Sir John F. W. Herschel, letter on "Le Verrier's planet," *Athenaeum* **1019** (October 3, 1846). The letter was written October 1, 1846.
12. J. W. Herschel to W. Lassell, October 1, 1846, Royal Society Herschel papers (H.S.22.285, quoted in R. W. Smith and R. Baum, "William Lassell and the Ring of Neptune: A Case Study in Instrumental Failure," *Journal for the History of Astronomy xv*, 1–17:2 (1984).
13. See Richard Baum, *The Planets: Some Myths and Realities*, pp. 120–146; also Baum and Robert W. Smith, "Neptune's Forgotten Ring," *Sky and Telescope* **77**, 610–611 (1989), and Smith and Baum, "William Lassell and the Ring of Neptune: A Case Study in Instrumental Failure," *Journal for the History of Astronomy xv*, 1–17 (1984).
14. Airy to Challis, October 14, 1846; quoted in Smart, "John Couch Adams," 65.
15. Airy to Le Verrier, October 14, 1846; quoted in Glaisher, Biographical Notice, Adams, *Scientific Papers*, vol. 1, p. xxiv.
16. Airy to Le Verrier, October 18, 1846; quoted in Grosser, *Discovery of Neptune*, 131–132.
17. *Comptes Rendu* **23**, 751 (1846).
18. Quoted in Harrison, *Voyager*, 33.
19. Ibid.
20. *L'Univers*, October 21, 1846.
21. *Le Semaine*, October 25, 1846.
22. F. Arago, "Examen des remarques critiques et des questions de priorité que la découverte de M. Le Verrier a soulevées," *Comptes Rendu* **23**, 751, 754 (1846).
23. Mary Roseveare to W. M. Smart, April 21, 1947; quoted in Harrison, *Voyager*, 71.
24. E. Loomis, "The Discovery of the Planet Neptune," in *Progress of Astronomy* (New York: Harper & Bros., 1851), 58–59.

25. Chapman, "Airy and Neptune," 135.

26. Herschel to Sheepshanks, Dec. 17, 1846; cited in Robert W. Smith, "Cambridge Network," 416 note.

27. Herschel to Rev. Richard Sheepshanks, Dec. 17, 1846; quoted in Smith, "Cambridge Network," 416. The Latin phrase, from Virgil is translated "the gods will otherwise."

28. Herschel to R. Jones, quoted in Smith, "Cambridge Network," 416–417.

29. J. Herschel, *Monthly Notices of the Royal Astronomical Society* **11,** 111 (1848).

30. Encke to Le Verrier, September 28, 1846; MS letter in Paris Observatory library, quoted in Grosser, *Discovery of Neptune,* 119.

31. Airy, "Account," 142.

32. Agnes Clerke, *History of Astronomy,* 102.

33. Hanson, 359.

34. Ibid., 363–364.

35. Loomis, 59.

36. Quoted in Phillipe de la Cortadière and Patrick Fuentes, *Camille Flammarion* (Paris: Flammarion, 1994), 50.

37. Hanson, "Zenith and Nadir," 360.

38. Pannekoek, *History of Astronomy,* 361–362.

39. Loomis, 59.

40. J. Challis, "Determination of the Orbit of the Planet Neptune," *Astronomische Nachrichten 26,* 309–314 (1847).

41. For Herschel's observation, see Dennis Rawlins, "The Unslandering of Sloppy Pierre," p. 26; for Galileo's, see Stillman Drake and Charles T. Kowal, "Galileo's Sighting of Neptune," *Scientific American* **243,** 74–79 (1980).

42. B. Peirce, "Investigation in the Action of Neptune to Uranus," *Proceedings of the American Academy of Arts and Sciences* **1,** 65 (1847).

43. J. Babinet, "Sur la position actuelle de la planete située au de la de Neptune, et provisoirement nommée Hypèrion," *Comptes Rendu* **27,** 202–208 (1848).

44. Sir John Herschel, *Outlines of Astronomy* (New York: Appleton, 1876 ed.), 549n.

45. Ibid., 550–551.

46. J. C. Adams, Appendix on the Discovery of Neptune, *Liouville's Journal de Mathématiques,* New Series, Tome II (1876); in Adams, *Scientific Works,* vol. 1, pp. 63–65. Adams's comments were in French and have been translated by the authors.

47. Loomis, 59.

Chapter 9

1. C. Flammarion, *Popular Astronomy,* trans. J. Ellard Gore (London: Chatto & Windus, 1894), 466.

2. Charles Joseph Etienne Wolf, *Histoire de l'Observatoire de Paris de sa fondation a 1793* (Paris: Gauthier-Villars, 1902), 194.

3. M. J. Bertrand, *L'Académie et les Académiciens,* de 1666 a 1793; quoted in ibid., 215.

4. For a vivid, but not balanced, appraisal of Airy's rule, see Joseph Ashbrook, "The Airy Regime at Greenwich," in *Astronomical Scrapbook* (Cambridge, Mass. Sky Publishing Corporation 1984), 41–47.

5. E. Dunkin, "M. Le Verrier," *The Observatory* **1,** 199–206:204 (1877).

6. Sources for Flammarion's life include his autobiographical *Mémoires biographiques et philosophiques d'un astronome* (Paris: E. Flammarion, 1911), Emile Touchet, "La Vie et L'Oeuvre de Camille Flammarion, *Bulletin Societé de Astronomique de France* **39,** 341–365 (1925), and Luigi Prestinenza, "Camille Flammarion," *L'Astronomia* **52,** 34–44 (1986); above all, see the very readable biography by Philippe de la Cotardière and Patrick Fuentes, *Camille Flammarion* (Paris: Flammarion, 1994).

7. Quoted in Cotardière and Fuentes, 51.

8. Ibid., 50.

9. Ibid.

10. Ibid., 56.

11. Samuel Johnson, *Lives of the English Poets: A Selection* [London: Dent (Everyman's Library)], 106.

12. J. C. Adams, "Address on Presenting the Gold Medal of the Royal Astronomical Society to M. Le Verrier," in *Scientific Papers*, vol. 1, p. 357.

13. U. J. J. Le Verrier, "Nouvelles recherches sur les mouvements des planètes," *Comptes Rendu* **29,** 1–3:2 (1849).

14. Jacques Laskar, "Appendix: The stability of the solar system from Laplace to the present," *in Planetary Astronomy from the Renaissance to the Rise of Astrophysics, Part B,* eds. Taton and Wilson, 245. As later shown by Henri Poincaré, the series approximations used by astronomers were in general divergent; thus they only served to represent the motions of the planets for a limited period of time. Since then, computers have been used to show that for the smaller planets, their motion becomes chaotic, preventing all prediction beyond about 100 million years. Or as Laskar says, the solar system is unstable (p. 247).

15. Zachary, 74.

16. Ibid., 81.

17. U. J. J. Le Verrier, "Lettre de M. Le Verrier à M. Faye sur la théorie de Mercure et sur le mouvement du périhélie de cette planète," *Comptes Rendu* **59,** 379 (1859).

18. U. J. J. Le Verrier, "Theorie du mouvement de Mercure," *Annales de l'Observatoire Impérial de Paris* (Mémoirs) **V,** 78 (1859). Hereafter "Theorie."

19. Ibid., 99.

20. Le Verrier, "Lettre à Faye," 381.

21. Ibid., 381.

22. Cited by Hanson, 369.

23. The authors are indebted to Professor Donald E. Osterbrock for the lucid explanation given here.

24. Le Verrier, "Theorie," 105. Hanson considers it possible that in the mid-1840s Le Verrier may have considered the existence of a sun-obscured planet, i.e., a planet in line with the Earth and Sun, hence never visible like the old antichthon of the Greeks. However, this straight-line configuration would be unstable, as demonstrated by Le Verrier's colleague J. Liouville, "Sur un cas particulier du problem des trois corps," *Comptes Rendu* **14,** 503–506 (1842). See Hanson, 368 and 377–378.

25. Le Verrier, "Theorie," 105.

26. Le Verrier, "Lettre à Faye,"383.

27. J. H. Schroeter, *Beobachtungen über die Sonnenfackeln und Sonnenflecken* (Erfurt: Georg Adam Keyser, 1789).

28. See "Extract of a Letter from M. Schwabe to Mr. Carrington," *Monthly Notices of the Royal Astronomical Society* **17,** 241 (1857); also "Address Delivered by the President, M. J. Johnson, Esq. on Presenting the Gold Medal of the Royal Society to M. Schwabe," *Monthly Notices of the Royal Astronomical Society* **17,** 126–131 (1857).

29. Schwabe, "Extract," 241.

30. Ibid.

31. Joseph Ashbrook, "Julius Schmidt: An Incredible Visual Observer," in *Astronomical Scrapbook,* 253.

32. E. C. Herrick, "Lettre de M. Herrick à M. Le Verrier," *Comptes Rendu* **49,** 810–812 (1859). See also Herrick, "Supposed Planet between Mercury and the Sun," *American Journal of Science,* Series 2, **28,** 445–446 (1859). Herrick notes p. 445, "In this connection it may be worthwhile to state that there are already on record observations which make it highly probable that there exists an intra-Mercurial planet with a satellite." He then quotes Wartmann p. 446, "that Pastorff, of Bucholz, an attentive observer of the solar spots, saw twice in 1836 and once in 1837 two round black spots of unequal size, moving across the sun, changing their place rapidly, and pursuing each time routes somewhat different." In 1834 Pastorff also claimed to have seen two small bodies, suggesting a planet and its satellite, pass across the Sun no less than six times; they required only a few hours for their transits. "They had the appearance . . . like that of Mercury in its transits" (p. 446).

33. R. Wolf, *Mittheilungen über die Sonnenflecken* **(10),** (1859).

34. T. Dick, *Celestial Scenery: Or, The Wonders of the Planetary System Displayed; Illustrating the Perfection of Deity and a Plurality of Worlds* (London: Thomas Ward & Co. 1838), pp. 279–280.

35. Remark attributed to Benjamin Peirce as having been made at the Three Hundred and Forty-first Meeting of the American Academy of Arts and Sciences, January 7, 1851. *Proceedings of the American Academy of Arts and Sciences* **II,** 251 (1852). See also ibid. IV, 411 (1854).

36. D. Kirkwood, "On a New Analogy in the Periods of Rotation of the Primary Planets," *American Journal of Science and Arts,* 2nd series., *9,* 395–399 (1850). For historical overview see R. L. Numbers, "The American Kepler: Daniel Kirkwood and His Analogy," *Journal for the History of Astronomy* **IV,** 13–21 (1973).

37. J. Babinet, "Mémoire sur les nuages ignes du soleil considérés comme des masses planétaires, *Comptes Rendu* **22,** 281–286 (1846).

38. "Lettre de M. Buys Ballot a M. Le Verrier," *Comptes Rendu* **49,** 812–813 (1859).

39. Tisserand, *Traité de mécanique céleste* (Paris, 1888–1896), vol. IV, preface.

40. Adams, "Address," p. 356.

Chapter 10

1. M. Quidet, Maire d'Orgères-en-Beauce, to Richard Baum, June 25, 1974; personal correspondence.

2. Rev. James Challis, "On the Planet Within the Orbit of Mercury, Discovered by M. Lescarbault," *Proceedings of the Cambridge Philosophical Society* **1,** 219–222:219 (1865).

3. John Milton, *Paradise Lost,* III, lines 588–590.

4. "Passage d'une Planète sur le disque du Soleil, observée à Orgères, par M. Lescarbault. Lettre à M. Le Verrier," *Comptes Rendu* **50,** 40–46 (1859); see also *Cosmos* **16,** 50ff (January 13, 1860).

5. Ibid., 44–45.

6. Ibid., 45.

7. David Brewster, "Recent Discoveries in Astronomy," *North British Review* **33,** 1–20:9 (1860).

8. Abbé Moigno, "Découverte d'une Nouvelle Planète entre Mercure et le Soleil," *Cosmos* **16,** 22ff (January 6, 1860).

9. The deal board on which Lescarbault made his calculations was later presented by Le Verrier to the Académie des Sciences.

10. "Passage d'une Plànete," 46.

11. Brewster, "Recent Discoveries," 12.

12. "Passage d'une Planète," 45–46.

13. Ibid., 46.

14. Brewster, "Recent Discoveries," 8.

15. Ibid., 12.

16. As noted in Robert Fontenrose, "In Search of Vulcan," *Journal for the History of Astronomy iv,* 145–158: 146 (1973).

17. "A Supposed New Interior Planet," *Monthly Notices of the Royal Astronomical Society* **20,** 98–100 (1860).

18. H. Tuttle, "Reminiscences of a Search for 'Vulcan' in 1860," *Popular Astronomy* **7,** 235–237:235–236 (1899).

19. Abbé Moigno, "Nouvelles de la Semaine," *Cosmos* **16,** 117 (February 3, 1860). Our thanks to the late Dr. E. W. Maddison, sometime librarian of the Royal Astronomical Society, for calling this to our attention.

20. Benjamin Scott, "The New Inferior Planet," to the Editor of the London *Times,* January 10, 1860. This and other documents on the 1847 observations are collected and reprinted in Rev. E. Ledger, *Intra-Mercurial Planets, a Lecture Delivered at Gresham College on Friday 14, 1879* (Cambridge: University Press, 1879).

21. Abbatt's scruples (or Scott's insistence) led him to pen a letter of testimonial, dated January 10, 1860: "I perfectly remember you mentioning to me that you had seen a new 'inferior planet' in the manner you have stated in to-day's *Times.* And I regret that I should have been the cause of preventing your making the fact more generally known." Abbatt's letter was published in the *Times* 2 days later.

22. Capel Lofft, "On the Appearance of an Opaque Body Traversing the Sun's Disc," *Monthly Magazine* 102–103 (March 1, 1818).

23. Agnes Clerke, *History of Astronomy,* 125.

24. Ibid.

25. R. C. Carrington, "List of Possible Transits of Inner Planets," *Monthly Notices of the Royal Astronomical Society* **20,** 100–101 (1860); also, "On some Previous Observations of Supposed Planetary Bodies in Transit over the Sun," *Monthly Notices of the Royal Astronomical Society* **20,** 192–194 (1860).

26. Joseph Sidebotham, "Note on an Observation of a Small Black Spot on the Sun's Disc," *Proceedings Literary and Philosophical Society of Manchester* **12,** 105 (1872–73).

27. Simon Newcomb, "On the Supposed Intra-Mercurial Planets," *Astronomical Journal* **6,** 162–163 (1860).

28. See J. Bauschinger, "Zur Frage über die Bewegung der Mercurperihelion," *Astronomiche Nachrichten* **109,** 27–32 (1884).

Chapter 11

1. R. Wolf, "Sur quelques Periodes qui semblent se rapporter, à les Passages de la Planète Lescarbault sur le Soleil," *Comptes Rendu* **1,** 482 (1860).

2. M. R. Radeau [sic], "Future Observations of the Supposed New Planet," *Monthly Notices of the Royal Astronomical Society* **20,** 195 (1860).

3. Reports from Ellery at Victoria Observatory, Scott at Sydney, and Tennant at Madras were published in *Monthly Notices of the Royal Astronomical Society* **20,** 344 (1860). A thorough search was undertaken by H. P. Tuttle, at the behest of G. P. Bond, director of the Harvard Observatory, throughout April 1860. "My instructions were," Tuttle later recalled, "to begin my work as soon after sunrise as possible and continue to observe the solar surface twice every hour until nearly sundown." Tuttle used a solar eyepiece and observed the Sun on 18 days during the month. He reported no planets seen, only ordinary sunspots; however, he noted, "what I did *see* with my right eye whenever I looked at a lighted lamp during the following three months, was all the colors of the solar spectrum! Had I continued these observations for a couple of months longer I should have ruined the sight of my right eye forever. . . ." See Tuttle, "Reminiscecenes of a Search for 'Vulcan' in 1860."

4. E. Liais, "Sur la Nouvelle Planète announcé par M. Lescarbault" (letter dated March 8, 1860) *Astronomische Nachrichten* **52,** 369 (no. 1248) (1860).

5. *L'espace céleste et la natur tropical* (Paris, 1866), 495.

6. Ibid., 498.

7. Liais, "Sur la Nouvelle Planète," 370.

8. Liais, *L'espace céleste,* 498.

9. Richard A. Proctor, *Myths and Marvels of Astronomy* (London: Chatto & Windus, 1878), 321.

10. Liais, *L'espace céleste,* 498.

11. Ibid., 499.

12. Ibid., 500.

13. "Recent Discoveries," 19–20.

14. W. F. Denning, "A Supposed New Planet," *Science for All* **4,** 264–270:267 (1893).

15. J. R. Hind, "Note on a Dark, Circular Spot upon the Sun's Disk, with Rapid Motion, as Observed by W. Lummis, Esq., of Manchester, 1862, March 20," *Monthly Notices of the Royal Astronomical Society* **22,** 232 (1862). Hind adds that "it is evident, from the sketch, that Mr. Lummis's estimate of the arc passed over during the twenty-two minutes he watched the spot is much too great. It would be nearer 6' than 12'."

16. J. R. Hind, letter, the London *Times,* October 19, 1862.

17. "Lettre de M. Le Verrier addressée a M. le Marechal Vaillant," *Comptes Rendu* **60,** 1113–1115 (1865).

18. "Minute Object Seen near the Sun," Report of Mr. W. S. Gilman, Jr., *Astronomical and Meteorological Observations of the U.S. Naval Observatory* (Washington, DC: United States Government Printing Office, 1870), 180.
19. Noted in W. H. DeShon, *Utica Morning Herald,* excerpted in *Hamilton Literary Monthly* (October 1876), 115.
20. Letter of B. A. Gould, Jr. to Yvon Villarçeau, September 7, 1869, *Comptes Rendu* **69,** 813–814 (1869).
21. J. R. Hind, "Stellar Objects Seen during the Eclipse of 1869," *Nature* **18,** 663–664 (1878).
22. See W. F. Denning's correspondence in the *Astronomical Register* **7,** 89, 113 (1869); **8,** 77–78, 108–109 (1870); **9,** 64 (1871).
23. As noted in Fontenrose, "In Search of Vulcan," 148.
24. J. C. Adams, "Address," 356.
25. Simon Newcomb, *The Reminiscences of an Astronomer* (New York: Houghton Mifflin Co., 1903), 328.
26. Willy Ley, *Watchers of the Skies* (New York: Viking, 1966), p. 198.
27. J. J. Thomson, quoted in Rupert T. Gould, *Oddities* (London: Geoffrey Bles, 1944), 196.
28. As noted in Émile Touchet, "La Vie et L'Oeuvre de Camille Flammarion," *Bulletin Société de Astronomique de France* **39,** 341–365 (1925).
29. C. Flammarion, *Popular Astronomy,* 346.
30. Undated clipping, Mary Lea Shane Archives of the Lick Observatory.
31. Newcomb, *Reminiscences,* 329.
32. Ibid.
33. Clerke, *History of Astronomy,* 219 (1885 ed.); 171 (1908 ed.).
34. Morando, "The Golden Age of Celestial Mechanics," 229.
35. U. J. J. Le Verrier to ?? November 2, 1874; in Mary Lea Shane Archives of the Lick Observatory. Someone has written "to G. P. Bond" on this letter. However, this cannot be correct; Bond had died in 1865. Almost certainly the letter was to E. S. Holden.
36. Adams, "Address," 358–359.
37. Le Verrier, "Théorie nouvelle du mouvement de la planète Neptune: Remarques sur l'ensemble des théories des huit planètes principales: Mercure, Vénus, la Terre, Mars, Jupiter, Saturne, Uranus et Neptune." *Comptes Rendu* **79,** 1424 (1874).
38. Wolf to Le Verrier, letters, August 26 and September 6, 1876, *Comptes Rendu* **83,** 510, 561 (1876).
39. Richard Proctor, "The Planet Vulcan," *English Mechanic and World of Science* **605,** 160 (October 27, 1876).
40. Wolf to Le Verrier, letters of August 26 and September 6, 1876, *Comptes Rendu* **83,** 510, 561 (1876).
41. Le Verrier to Wolf, September 12, 1876; in R. Wolf, *Astronomische Mittheilungen* **26,** 377–378 (1881).
42. Le Verrier to Wolf, September 21, 1876; ibid.
43. Le Verrier to Wolf, September 29, 1876; ibid.
44. W. H. De Shon, in *Hamilton Literary Monthly* **115** (October 1876).
45. Sir G. B. Airy, "Note on the Sunspot of April 4, 1876" (Telegram of October 4, 1876), *Nature* **14,** 534 (1876).

46. Proctor, *Myths and Marvels* (1878), 309–326:321.

47. Lewis Swift to E. E. Barnard, July 13, 1881; Vanderbilt University Archives.

48. "The Inter-Mercurial Planet," *Scientific American* (October 21, 1876), 257.

49. Fontenrose, "In Search of Vulcan," 149.

Chapter 12

1. U. J. J. Le Verrier, "Les planètes intra-mercurielles" (suite), *Comptes Rendus* **83**, 719 (1876).

2. Airy, "Account," 142.

3. "Vulcan Again," *Scientific American* (November 18, 1876), 321.

4. Hanson, 376.

5. Ibid.

6. Proctor, *Myths and Marvels*, 326.

7. As noted in an obituary of Le Verrier in the London *Times*, October 11, 1877.

8. E. Dunkin, "Le Verrier," *The Observatory* **1**, 206 (1877).

Chapter 13

1. Sources used for the life of James Craig Watson are: George C. Comstock, "Memoir of James Craig Watson, 1838–1880," *National Academy of Sciences Biographical Memoirs*, vol. III (Washington, D.C., 1895), pp. 45–57; Heber D. Curtis, "James Craig Watson, 1838–1880," *Michigan Alumnus Quarterly Review* **Summer** (1938).

2. J. C. Freeman, "Prof. Jas. C. Watson," *Milwaukee Sentinel* (June 23, 1880).

3. Comstock, "Memoir," 46.

4. Freeman, "Prof. Jas. C. Watson."

5. Comstock, "Memoir," 47.

6. Ibid., 49.

7. Freeman, "Prof. Jas. C. Watson."

8. "Dr Peters' 'Eunike'—his experience in Constantinople," undated clipping in scrapbook in Mary Lea Shane Archives of the Lick Observatory.

9. See Joseph Ashbrook, "The Adventures of C. H. F. Peters," in *The Astronomical Scrapbook*, 56–66, which contains details of the most famous lawsuit that involved Peters: the case of Peters v. Borst.

10. As noted in Gary W. Kronk, *Comets: A Descriptive Catalog* (Hillside, New Jersey: Enslow Press, 1984), 47. By longstanding tradition, comets are, with a few notable exceptions such as Halley's, Lexell's, and Encke's comets, named for their discoverers.

11. "Dr Peters' 'Eunike'," A letter from C. A. Young to G. C. Comstock, August 22, 1887, contains some additional details about the discovery. Young points out that Watson had his (and also some of the French astronomer Chacornac's) star charts with him. "As to Juewa," writes Young, "he *did* say that the minute he looked into the field, and before he had examined the chart with any care, he at once recognized the presence of an interloper by his memory of the field alone. . . . But he had a chart of the field, and used it. The field had a certain peculiar configuration in it . . . such that an interloper was made pretty conspicuous by it [and] I think he had used the very stars as

comparison stars in the obs[ervatio]n of some planet within a year or two, so that the field was specially impressed upon his memory." From University of Wisconsin Archives, Department of Astronomy Records, George C. Comstock Papers.

12. Quoted by W. H. DeShon, *Hamilton Literary Monthly* (October 1876), p. 115.

Chapter 14

1. John A. Eddy, "The Great Eclipse of 1878," *Sky and Telescope* **45,** 340 (1973).

2. C. H. F. Peters to E. S. Holden, June 8, 1878; Mary Lea Shane Archives of the Lick Observatory.

3. E. S. Holden, "Reports on the Total Solar Eclipses, July 29, 1878 and January 11, 1880," *Astronomical and Meteorological Observations Made During the Year 1876 at the United States Naval Observatory,* pt. II, Appendix III (Washington, D.C., 1880), 145.

4. William Harkness, ibid., 30.

5. J. A. Eddy, "Thomas A Edison and Infra-red Astronomy," *Journal for the History of Astronomy* **3,** 165–187 (1972).

6. Freeman, "Prof. Jas. C. Watson."

7. D. P. Todd, "Preliminary Account of a Speculative and Practical Search for a Trans-Neptunian Planet," *American Journal of Science* **20,** 225–234 (1880).

8. J. Norman Lockyer, "The Eclipse," *Nature* **18,** 461 (August 29, 1878).

9. James C. Watson, report in *Washington Observations,* 1876, Appendix III, 117.

10. W. T. Sampson, ibid., 109.

11. *Crofutt's New Overland Tourist and Pacific Coast Guide,* (Omaha, Neb. and Denver, Col.: Overland Publishing Co., 1882), 68.

12. Eddy, "Great Eclipse," 344.

13. J. A. Eddy uncovered the brick piers during his excavations in 1968 and 1973.

14. Eddy, "Great Eclipse," 341.

15. E. L. Trouvelot, *Washington Observations,* 1876, Appendix III, 75.

16. S. Newcomb, ibid., 102.

17. Watson, ibid., 118.

18. Lockyer, "The Eclipse," *Nature* **18,** 457–462:462 (August 29, 1878).

19. Watson, *Washington Observations,* 1876, Apendix III, 117.

20. Donald E. Osterbrock, "Lick Observatory Solar Eclipse Expeditions," *The Astronomy Quarterly* **3,** 70 (1980).

21. Newcomb, *Washington Observations,* 1876, Apendix III, 104.

22. Ibid., 105.

23. Watson, ibid., 119.

24. Ibid., 120.

25. Newcomb, ibid., 105.

26. G. W. Hill, "Biographical Memoir of Asaph Hall," *Biographical Memoirs, National Academy of Sciences* **6,** 240–275:264 (1908).

27. *Nature* **18,** 426 (August 15, 1878). Todd appreciated the value of rapid communication and subsequently wrote two papers on the subject "On the Use of the Electric Telegraph During Total Solar Eclipses" *Proceedings of the American Academy of Arts and Sciences* **16,** 359–363 (1881), and "On Observations of the Eclipse of 1887, Aug. 18 in Connection with the Electric Telegraph," *American Journal of Science* **33,** II series 226–228 (1887).

Chapter 15

1. Quoted in Eddy, "Great Eclipse," 345.
2. *Nature* **18,** 380–381 (August 8, 1878).
3. Ibid., 385.
4. Ibid.
5. Daniel Kirkwood, "The Planet Vulcan," *The Popular Science Monthly* **13,** 732–735:735 (1878). For earlier pronouncements by Kirkwood, see "On the Probable Existence of Undiscovered Planets," *Literary Record and Journal of the Linnaean Association of Pennsylvania College* **3,** 131 (1847), "On a New Analogy in the Periods of the Primary Planets," *American Journal of Science and the Arts,* Ser. 2, **9,** 395–399 (1850), "On Certain Analogies in the Solar System," ibid. **14,** 210–219 (1852), and "On Certain Harmonies of the Solar System," ibid. **38,** 1–18 (1864).
6. Lewis Swift, report in *Washington Observations,* 1876 Appendix III, 226.
7. *Nature* **18,** 433 (August 22, 1878) specified as originally in *Rochester Democrat,* reprinted in *New York Tribune.*
8. Ibid.
9. Ibid. See also, "Letter from Mr. Lewis Swift, Relating to the Discovery of Intra-Mercurial Planets," *American Journal of Science and Arts,* 3rd Series, **16,** 313–315 (1878).
10. *Nature* **18,** 495 (September 5, 1878).
11. Ibid.
12. Ibid., p. 496. Of four possible orbits proposed by Le Verrier, Gaillot found that the one agreeing most closely with Watson's position had a semimajor axis of 0.164, period of revolution of 24.25 days—less than the period of the Sun's rotation—and an eccentricity of 0.14, comparable to that of Mercury. The longitude of the perihelion was 74°. With regard to the inclination of the orbit to the ecliptic, Gaillot found it could not exceed 7°. The most serious objection to the identification of the object observed by Watson with that indicated by Le Verrier's formula was that "we should see but a very small part of the disc illuminated, and without denying that there is reason in this objection, M. Gaillot adds that Prof. Watson describes 'as being of the fourth magnitude, a star the diameter of which may be comparable with that of Mercury, and which, in superior conjunction, may appear of the first magnitude.' He further remarks while it is not possible to decide with certainty upon the identity of Prof. Watson's planet with that of which Le Verrier has indicated the track, he believes he has shown that there is no incompatibility between the observed and hypothetical objects. If only one such planet exists between Mercury and the Sun, M. Gaillot points out that, in order to account for the accelerated motion in the perihelion of Mercury, its mass must be nearly equal to that of the latter."
13. E. Mouchez, "Nouvelle observation probable de la planète Vulcan par M. le professeur Watson," *Comptes Rendus* **87,** 229 (1878).
14. *New York Times,* August 16, 1878.
15. *Nature* **18,** 495–496 (Sept. 5, 1878). The position for his second object Watson gave, for July 29, 1878, 5h 17m 46s Washington Mean Time: R. A. 8h 8m 38s, declination 118° 3'.
16. J. C. Watson, "On the Intra Mercurial Planets; from letters to the editors, dated Ann Arbor, Sept. 3d, 5th and 17th, 1878," *American Journal of Science and Arts,* 3rd Ser. **16,** 310–313:311 (1878).

17. Asaph Hall, report in *Washington Observations,* 1876 Appendix III, 254.

18. Watson, "On the Intra-Mercurial Planets," 312.

19. Ibid.

20. Lewis Swift, letter September 4, 1878, *Nature* **18,** 539 (September 19, 1878).

21. Ibid.

22. Swift, "The Intra-Mercurial Planets," *Nature* **19,** 96 (Dec. 5, 1878).

23. J. C. Watson, "Schreiben des Herrn Prof. Watson an den Herausgeber," *Astronomische Nachrichten* **93** (2217), 141–142 (1878); **93** (2220), 189–192 (1878); **93** (2223), 239–240 (1878).

24. C. A. F. Peters to James C. Watson, September 24, 1878; University of Wisconsin archives.

25. Isaac H. Hall, Memorial Address, Hamilton College, 1896.

26. C. H. F. Peters, "Some Critical Remarks on So-called Intra-mercurial Planet Observations," *Astronomische Nachrichten* **94** (2253), 321–340 (1879).

27. Joseph Ashbrook, "The Adventures of C. H. F. Peters," *Astronomical Scrapbook,* (Cambridge: Sky Publishing Corp., 1984), 63.

28. Peters, "Critical remarks," *Astronomische Nachrichten* **94** cols. 323–324 (1879).

29. Ibid., cols. 325–326.

30. Ibid., col. 326.

31. Ibid., cols. 327–328.

32. Ibid., cols. 335–336.

33. Agnes M. Clerke, *History of Astronomy* (London: Adam & Charles Black, 1908), p. 250.

34. C. A. Young to C. H. F. Peters, June 2, 1879, Hamilton College Library, Clinton, New York.

35. J. C. Watson, "Schreiben des Herrn Prof. Watson an den Herausgeber," *Astronomische Nachrichten* **95** cols. 101–106:103–104 (1879).

36. Ibid., col. 104.

37. *Wisconsin State Journal,* September 18, 1877, 7.

38. "The Washburn Observatory," *The Observatory* **6,** 280–281 (1883); R. C. Bless, *Washburn Observatory 1878–1978* (Madison: University of Wisconsin-Madison, 1978).

39. *Wisconsin State Journal,* September 28, 1878.

40. Ibid., October 11, 1878.

41. J. C. Watson, "Report of the Director of the Observatory and Professor of Astronomy to the Board of Regents," *Annual Report of the Regents of the University of Wisconsin* 36 (1880).

Chapter 16

1. Theodor von Oppolzer, "Sur l'existence de la planète intra-mercurielle indiquée par Le Verrier," *Comptes Rendu* **88,** 26–27 (1879) and "Elemente des Vulkan," *Astronomische Nachrichten* **94** cols. 97–100 (1879).

2. C. H. F. Peters, "Schreiben des Herrn Prof. C. H. F. Peters an den Herausgeber," *Astronomische Nachrichten* **94** cols. 303–304 (1879).

3. Oppolzer, "Bemerkung zu dem Aufsatze: 'Elemente des Vulkan," *Astronomische Nachrichten* **94** cols. 303–304 (1879).

4. Clippings book, Mary Lea Shane Archives of the Lick Observatory.

5. Edgar Frisby, report on eclipse of January 11, 1880; in *Washington Observations,* 1876 Appendix III, 395–410.

6. Clippings book, Mary Lea Shane Archives of the Lick Observatory.

7. Peters to the Editor of the *Utica Morning Herald*, Feb. 8, 1880; clippings book, Mary Lea Shane Archives of the Lick Observatory.

8. James C. Watson, "The Problematical Vulcan," letter to the editor of the *Madison (Wisconsin) State Journal* (February 18, 1880); clippings book, Mary Lea Shane Archives of the Lick Observatory.

9. Ibid.

10. Humboldt, in *Cosmos* **3**, 72–73 (1851) and footnote, declares: "The question whether stars can be seen in daylight with the naked eye through the shafts of mines, and on very high mountains, has been with me a subject of inquiry since my early youth. I was aware that Aristotle had maintained that stars might occasionally be seen from caverns and cisterns, as through tubes. Pliny alludes to the same circumstance, and mentions the stars that have been most distinctly recognized during solar eclipses. The chimney-sweepers whom I have questioned agree tolerably well in the statement that 'they have never seen stars by day, but that, when observed at night, through deep shafts, the sky appeared quite near, and the stars larger.' " Aristotle's reference appears where it is least expected in *De Generat. Animal.*, V. i; "keenness of sight," he says there, "is as much the power of seeing far, as of accurately distinguishing the differences presented by the objects viewed. These two properties are not met with in the same individuals. For he who holds his hand over his eyes, or looks through a tube, is not on that account more or less able to distinguish differences of color, although he will see objects at a greater distance. Hence it arises that persons in caverns or cisterns are occasionally enabled to see stars."

11. This description follows that given by E. S. Holden in *Vierteljahreschrift der Astronomischen Gesellschaft* **18,** 112–116 (1883).

12. C. A. Young to G. C. Comstock, August 22, 1887; University of Wisconsin Archives, George C. Comstock Papers.

13. Apparently his ill treatment of his wife did not begin after his death. According to C. A. Young, writing to G. C. Comstock, "his treatment of his wife was simply abominable. She was rather weak and querulous, ... Of course I do not expect or desire this estimate of his character to go into the narrative of his life. I give it only to forestall any extravagant laudations which would vitiate the picture entirely for any one who knew the subject a little intimately. There is no need to expose his faults; but they should not be replaced by virtues he did not possess." Further on Watson's character, Young remarked that "he was one of the most energetic and able men I ever knew . . . extremely self-confident (but not perhaps more so than his abilities justified) selfish and unscrupulous in advancing his own interests." C. A. Young to G. C. Comstock, August 22, 1887; University of Wisconsin Archives, George C. Comstock Papers.

14. Comstock, *James Craig Watson*. p. 55. He was buried at Oakwood Cemetery, Ann Arbor. Watson's asteroids and the dates of their discovery are as follows:

No.	Name	Date
79	Eurynome	September 14, 1863
93	Minerva	August 24, 1867
94	Aurora	September 6, 1867
100	Hekate	July 11, 1868

101	Helena	August 15, 1868
103	Hera	September 7, 1868
104	Klymene	September 13, 1868
105	Artemis	September 16, 1868
106	Dione	October 10, 1868
115	Thyra	August 6, 1871
119	Altaea	April 3, 1872
121	Hermione	May 12, 1872
128	Nemesis	November 25, 1872
132	Aethra	June 13, 1873
133	Cyrene	August 16, 1873
139	Juewa	October 10, 1874
150	Nuwa	October 18, 1875
161	Athor	April 16, 1876
168	Sibylla	September 28, 1876
174	Phaedra	September 2, 1877
175	Andromache	October 1, 1877
179	Klytaemnestra	November 11, 1877

15. C. H. F. Peters to E. S. Holden, February 17, 1881; Mary Lea Shane Archives of the Lick Observatory.

16. E. S. Holden, "Report of the Director of the Observatory and Professor of Astronomy to the Board of Regents," *Annual Report of the Regents of the University of Wisconsin* **36** (1882).

17. According to the recollections of Albert Whitford, who worked under Joel Stebbins at the Washburn Observatory from 1931 until Whitford left, on leave to work on radar in the Radiation Laboratory at MIT in 1940.

18. It was located on the south slope of Observatory Hill, just east of Agricultural Hall. During a visit to Washburn Observatory in June 1996, Sheehan found nothing left of it; the place where it stood is now occupied by an asphalt-covered playground for the University of Wisconsin's Laboratory School for kindergarten children.

19. Lewis Swift to E. E. Barnard, dated "Rochester Declaration Day," 1882; Vanderbilt University archives. Without further positional data, no orbit could be calculated, and the comet has never received an official designation.

20. Swift to Barnard, February 10, 1883; Vanderbilt University archives.

21. Lewis Swift, "The Intra-Mercurial Planet Question Not Settled," *Sidereal Messenger* **3**, 242–244 (1883–84).

22. E. S. Holden, "Report on the Eclipse of May 6, 1883," *Memoirs of the National Academy of Sciences* **2**, 100–102 (1884).

23. Fontenrose, "In Search of Vulcan," 153.

24. Isaac H. Hall, Memorial Address, 27. Peters remained active to the very last and had discovered his 48th—and last—new asteroid, 287 Nephthys, in 1889, when he was 76 years old.

25. Lewis Swift, "The Intra-Mercurial Planet Question." Following the 1883 eclipse he wrote (p. 242): "Scientific journals both in this country and in Europe appear to entertain the idea that because no intra-Mercurial planet during the last total eclipse was seen, none therefore exists. . . . As well might I assert that there was no comet where Brooks discovered that of 1812, because on three different occasions only a few days previously, I searched thoroughly over the region and found none." After recalling his own observations of the two objects at the eclipse of 1878 and Trouvelot's ruddy suspicious object, he added (p. 244): "My faith in their existence was never stronger than today."

26. Richard A. Proctor, *Old and New Astronomy* (London: Longmans & Green, 1892), 427. Proctor, however, had died in 1888.

27. Simon Newcomb, "Discussion and Results of Observations on Transits of Mercury from 1677 to 1881," *Astronomical Papers of the American Ephemeris and Nautical Almanac* **1**, 367–487 (1882).

28. Newcomb, "The Elements of the Four Inner Planets and the Fundamental Constants of Astronomy," *Supplement to the American Ephemeris and Nautical Almanac 1897* (Washington, DC, 1895).

29. Hanson, 374.

30. A. Hall, "A Suggestion in the Theory of Mercury," *Astronomical Journal* **14**, 49–51 (1894).

31. Hanson, 376–377.

32. Lewis Swift, "Intra-Mercurial Planets," *English Mechanic and World of Science* **xlix,** 135 (September 25, 1896).

33. Lewis Swift to E. E. Barnard, February 27, 1882; Vanderbilt University archives.

34. Swift, "Intra-Mercurial Planets."

35. Ibid.

36. "Observation d'une étoile d'un éclat comparable à celui de Régulus et située dans la même constellation," *Comptes Rendu* **112,** 152–153 (1891).

37. Ibid., p. 260, C. note by Flammarion.

38. "FRAS," *English Mechanic and World of Science* **xliv,** 182–183:182 (October 9, 1896).

39. P. L. Brown, *Comets, Meteorites and Men* (London: Robert Hale & Co., 1973), p. 118.

40. L. Swift, "Note by Dr. L. Swift," *Astronomical Journal* **17,** 8 (1896).

41. L. Swift, "Were They Comets?" *English Mechanic and World of Science 58,* 1–61:36 (1898).

42. "Total Solar Eclipse of 1898, Jan. 22. Preliminary Reports on the British Government Expeditions," *Monthly Notices of the Royal Astronomical Society* (Appendix) *58,* 1–61:36 (1898).

43. A. Hall, "Plus Probans Quam Necesse Est," *Popular Astronomy* **7,** 13 (1899).

44. W. F. Denning, "Search for an Intra-Mercurial Planet," *Knowledge* **23,** 134 (1900).

45. Described by Pickering's brother, Edward C. Pickering, "A Photographic Search for an Intramercurial Planet," *Harvard Circular* **48** (February 13, 1900); reprinted in *Astrophysical Journal* **11,** 322–324 (1900).

46. C. G. Abbot, "A Preliminary Report of the Smithsonian Astrophysical Observatory Eclipse Expedition of May, 1900," *Report of the Smithsonian Astrophysical Observatory 1891–1901, to the 57th Congress of the USA* (Document #20, Exhibit D). 295–308: 308 (Washington D.C. 1902).

47. Perrine's career and searches for Vulcan are well described in Donald E. Osterbrock, John R. Gustafson, and W. J. Shiloh Unruh, *Eye on the Sky: Lick Observatory's First Century* (Berkeley: University of California Press, 1988), pp. 161–163.

48. W. W. Campbell, "The Closing of a Famous Astronomical Problem," *Popular Science Monthly* 494–503:500 (1909).

49. A. Einstein, *The Theory of Relativity* (London, 1924), 103.

50. Steven Weinberg, *Gravitation and Cosmology* (New York: Wiley, 1972), p. 198, gives a value of 42".11 plus or minus 0".45.

51. A. Einstein to P. Ehrenfest, January 17, 1916; quoted in Abraham Pais, "*Subtle is the Lord . . .*" (Oxford: Oxford University Press, 1982), p. 253. For technical details of the various theories to account for the anomalous advance of the perihelion of Mercury, including the successive ideas of Einstein on the subject, the reader is referred to the masterly account of N. T. Roseveare, *Mercury's Perihelion from Le Verrier to Einstein* (Oxford: Clarendon Press, 1982). Einstein's theory has now, of course, been abundantly confirmed except in the case of the binary star DI Herculis.

52. That year, S. Chandrasekhar could claim that "the only crucial empirical evidence for the aesthetically most satisfying physical theory conceived by the mind of man—Einstein's general theory of relativity—[was] the astronomical one derived from the motion of Mercury." S. Chandrasekhar, "The Case for Astronomy," *Proceedings of the American Philosophical Society* **108**, 1 (1964).

53. See R. A. Hulse and J. H. Taylor, "Discovery of a Pulsar in a Binary System," *Astrophysical Journal* **195**, L51–L53 (January 15, 1975). For this work Hulse and Taylor were awarded the 1993 Nobel Prize for Physics. The observations of this strange binary system over a period of twenty years have now provided a remarkably stringent test of General Theory which has proven correct to within one part in 10^{14}.

54. Curtis, "James Craig Watson."

55. Ibid.

Chapter 17

1. A. O. Leuschner, "Research Surveys of the Orbits and Perturbations of Minor Planets 1 to 1091 from 1801.0 to 1929.5." *Publications of the Lick Observatory* **XIX**, 369–370 (1935). Contributions of the Berkeley Astronomical Department (Students' Observatory), University of California.

2. C. V. L. Charlier, "Das Bodesche Gesetz und die sogenannten intramerkuriellen Planeten," *Astronomische Nachrichten* **193**, 269–272 (1912).

3. H. Block, *Arkiv for Mathem. Astronom. & Physik* **10** (1914).

4. R. Trumpler, "Search for Small Planets at the Triangle Points of Mercury and the Sun," *Publications of the Astronomical Society of the Pacific* **35**, 313–318 (1923).

5. Ibid., p. 316.

6. Ibid.

7. Ibid., p. 317.

8. W. M. Smart, "On the Motion of the Perihelion of Mercury," *Monthly Notices of the Royal Astronomical Society 82*, 12–19:12 (1921).

9. Ibid., pp. 12–13.

10. Ibid., p. 19.

11. Ibid.

12. Trumpler, pp. 317–318.

13. W. W. Campbell and R. Trumpler, "Search for Intramercurial Bodies," *Publications of the Astronomical Society of the Pacific* **35,** 214–215 (1923).

14. Ibid., 215.

15. Ibid., 216.

16. "The Mythical Planet Vulcan," *Science—Supplement* **75,** 8–9 (1932).

17. E. Huntington, *Earth and Sun: An hypothesis of Weather and Sunspots* (New Haven: Yale University Press, 1923).

18. H. C. Courten, "Evidence for Intramercurial Planetary Objects or Comets Discovered during Solar Eclipses," Preprint, Courten to Baum June 1971.

19. *International Astronomical Union Circular* **2558** (1973 July 12). See also F. Dossin, *Compte Rendu* **257,** 2246 (1973).

20. F. Tisserand, "Notice sur les planètes intra-mercurielles," *Annuaire Bureau des Longitudes 1882* 729–772:770.

21. K. Brecher, et al., "Is There a Ring around the Sun?" *Nature* **282,** 50–52 (November 1, 1979).

22. M. E. Bailey, S. V. M. Clubbe, and W. M. Napier, *The Origin of Comets* (Oxford: Pergamon Press, 1990), pp. 300–302.

Epilogue

1. E. S. Holden to D. O. Mills, February 6, 1888; Mary Lea Shane Archives of the Lick Observatory.

2. W. H. Pickering, "A Search for a Planet Beyond Neptune," *Annals of the Astronomical Observatory of Harvard College* **61,** 113 (1909); see also W. G. Hoyt, "William Henry Pickering's Planetary Predictions and the Discovery of Pluto," *Isis* **67,** 551–564 (1976).

3. J. B. A. Gaillot, "Tables nouvelle des mouvements d'Uranus et de Neptune," *Annales de l'Observatorie de Paris* **28** (1909).

4. J. B. A. Gaillot, "Contribution à la recherche des planètes ultra-neptuniennes," *Compte Rendu* **148,** 754–758 (1909).

5. W. G. Hoyt, *Planets X and Pluto* (Tucson: University of Arizona Press, 1980), p. 85.

6. P. Lowell, "Memoir on a Trans-Neptunian Planet," *Memoirs of the Lowell Observatory* **1,** 8 (1915).

7. Ibid., 101.

8. Clyde Tombaugh to William Sheehan, personal correspondence, April 27, 1990.

9. Mark Littmann, "Where is Planet X?" *Sky and Telescope* **78,** 596–599 (December 1989). Also Littmann, *Planets Beyond: Discovering the Outer Solar System* (New York: Wiley, 1990), contains a discussion of these recent efforts. Harrington, by the way, assumed that Tombaugh could not have missed Planet X, so he concentrated his searches in the southern Milky Way below Tombaugh's declination cut-off.

10. E. Myles Standish, Jr., "Planet X: No Dynamical Evidence in the Optical Observations," *Astronomical Journal* **105,** 2000–2006:2005 (1995).

11. Thomas Van Flandern to William Sheehan, personal correspondence, May 1, 1996, writes: "My understanding is that Standish has not removed the zonal systematic errors from his data. He considered the observational accuracy too

poor to be worth the effort. But the data we at the U.S. Naval Observatory analyzed was first carefully corrected zone by zone by determining the actual average errors in the 19th-century star positions using modern star observations extrapolated backwards, then applying these average errors to correct Uranus (and also Jupiter, Saturn, and Neptune, which also show systematic trends). Finally, many observations within short time spans were averaged to form normal points, in which the scatter due to random error was pretty minimal, and systematic trends stood out." Unfortunately, Van Flandern finds the residuals do not thus far allow a unique solution for the position of an unknown planet."

12. For a useful review, see Jane X. Luu and David C. Jewitt, "The Kuiper Belt," *Scientific American* **274,** 46–53 (May 1996).

Select Bibliography

A Note on Sources

Although not intended as an academic history, our work is nevertheless partially reliant on primary material held in the following places: Mary Lea Shane Archives of the Lick Observatory; Vanderbilt University; University of Wisconsin; Yerkes Observatory; Royal Astronomical Society; Royal Society; and the Paris Observatory. There is also a small cache of Lescarbault papers at Chateaudun, France, donated to the municipal library in 1928 by his daughter Mde. Guimberteau. In addition, a telescope belonging to Lescarbault is held by the museum at Chateaudun.

BOOKS

Alexander, A. F. O'D. *The Planet Uranus, A History of Observation, Theory and Discovery* (London: Faber & Faber, 1955).

Ashbrook, J. *The Astronomical Scrapbook, Skywatchers, Pioneers, and Seekers in Astronomy*, ed. L. J. Robinson (Sky Publishing Corp. and Cambridge University Press, 1984).

Bauschinger, J. *Untersuchungen ueber die bewegung des planeten Merkur* (Munich, 1881).

Bell, E. T. *Men of Mathematics* (New York: Simon and Schuster, 1937, reprinted 1986).

Berry, A. *A Short History of Astronomy from the Earliest Times Through the Nineteenth Century* (London: John Murray, 1898; New York: Dover, reprint 1961).

Caspar, M. *Kepler*, trans. and ed. by C. D. Hellman (London and New York: Abelard-Schuman, 1959). Annotated softcover edition with new introduction and references by O. Gingerich (New York: Dover, 1993).

Clerke, A. M. *A Popular History of Astronomy during the Nineteenth Century*, 4th ed. (London: A & C Black, 1908).

Crowe, M. *Theories of the World from Antiquity to the Copernican Revolution* (New York: Dover, 1990).

Crowe, M. *Modern Theories of the Universe from Herschel to Hubble.* (New York: Dover, 1994.)

Drake, Stillman. *Galileo: Pioneer Scientist* (Toronto: University of Toronto Press, 1990).

Dick, T. *Celestial Scenery: Or, the Wonders of the Planetary System Displayed; Illustrating the Perfection of Deity and a Plurality of Worlds* (London: Thomas Ward and Co., 1838).

Fauvel, J., Flood, R., Shortland, M., and Wilson, R., eds. *Let Newton Be!* (Oxford: Oxford University Press, 1988).

Galilei, Galileo. *Sidereus Nuncius or The Sidereal Messenger*, trans. by Albert van Helden (Chicago: The University of Chicago Press, 1989).

Gingerich, O. *The Eye of Heaven, Ptolemy, Copernicus, Kepler* (New York: The American Institute of Physics, 1993).

Gould, R. T. *Oddities; A Book of Unexplained Facts* (London: Geoffrey Bles, 1944), Chapt. X, "The Planet Vulcan."

Grant, E. *Planets, Stars & Orbs: The Medieval Cosmos, 1200–1687* (Cambridge: Cambridge University Press, 1994).

Grant, R. *History of Physical Astronomy from the Earliest Ages to the Middle of the Nineteenth Century* (London: Henry G. Bohn, 1852; New York: Johnson Reprint Corporation edition, 1966).

Grosser, M. *The Discovery of Neptune* (Cambridge: Harvard University Press, 1962).

Harrison, H. M. *Voyager in Time and Space, The Life of John Couch Adams, Cambridge Astronomer* (Sussex: The Book Guild, 1994).

Heath, T. L. *Aristarchus of Samos, The Ancient Copernicus, A History of Greek Astronomy to Aristarchus Together with Aristarchus's Treatise on the Sizes and Distances of the Sun and Moon* (Oxford: Clarendon Press, 1913; New York: Dover, reprint 1981).

Heath, Thomas L. *Greek Astronomy* (London: J. M. Dent & Sons Ltd., 1932; New York: Dover, reprint 1991).

Hoyt, W. G. *Planets X and Pluto* (Tucson: The University of Arizona Press, 1980).

Hunt, G. ed. *Uranus and the Outer Planets. Proceedings of the IAU/RAS Colloquium No 60* (Cambridge: Cambridge University Press, 1982).

Ley, W. *Watchers of the Skies, An Informal History of Astronomy from Babylon to the Space Age* (London: Sidgwick and Jackson, Ltd., 1964).

Liais, E. *L'Espace Céleste et la nature tropicale: Description Physique de l'Univers* (Paris: Garnier Frères, 1866).

Littmann, M. *Planets Beyond, Discovering the Outer Solar System* (New York: John Wiley and Sons Inc., 1988).

Loomis, E. *The Recent Progress of Astronomy Especially in the United States* (New York: Harper & Brothers, 1851; New York: Arno Press, reprint 1980).

Lubbock, C. A. *The Herschel Chronicle, The Life-Story of William Herschel and His Sister Caroline Herschel* (Cambridge: Cambridge University Press, 1933).

Kepler, J. *New Astronomy* trans. by W. H. Donahue (Cambridge: Cambridge University Press, 1992).

Koestler, A. *The Sleepwalkers, a History of Man's Changing Vision of the Universe* (London: Hutchinson, 1959).

Mitchell, O. M. *The Orbs of Heaven, or the Planetary and Stellar Worlds. A Popular Exposition of the Great Discoveries and Theories of Modern Astronomy* (London: G. Routledge & Co., 1857).

Moore, P. *The Planet Neptune* (Chichester: Ellis Horwood Ltd., 1988).

Newcomb, S. *The Reminiscences of an Astronomer* (Boston: Houghton & Mifflin, 1903).

Newton, I. *The Correspondence of Isaac Newton*, ed. by H. W. Turnbull (Vols. I-III), J. F. Scott (Vol. IV), and A. R. Hall and L. Tilling (Vols. V-VII) (Cambridge: Cambridge University Press, 1959–1977).

Newton, I. *The Mathematical Papers of Isaac Newton*, ed. by D. T. Whiteside (Cambridge: Cambridge University Press, 1967–1984).

Osterbrock, D. E. *James E. Keeler: Pioneer American Astrophysicist and the Early Development of American Astrophysics* (Cambridge: Cambridge University Press, 1984).

Osterbrock, D. E., Gustafson, J. R., and Shiloh Unruh, W. J. *Eye on the Sky, Lick Observatory's First Century* (Berkeley: University of California Press, 1988).

Pais, A. *"Subtle is the Lord . . ."* (Oxford: Oxford University Press, 1982).

Pannekoek, A. *A History of Astronomy* (London: George Allen & Unwin Ltd., 1961; New York: Dover reprint, 1989).

Poor, C. L. *The Solar System* (London: John Murray, 1908).

Proctor, R. A. *Myths and Marvels of Astronomy* (London: Chatto and Windus, 1878).

Proctor, R. A. *Old and New Astronomy*, completed by A. C. Ranyard (London: Longmans and Green, 1892).

Proctor, R. A. "New Planets Near the Sun." *In Rough Ways Made Smooth, A Series of Familiar Essays on Scientific Subjects.* (London: Longmans & Green, 1893), pp. 32–57.

Ronan, C. A. *Edmond Halley, Genius in Eclipse* (London: Macdonald, 1969).

Roseveare, N. T. *Mercury's Perihelion from Le Verrier to Einstein,* (Oxford: Clarendon Press, 1982).

Sarton, G. *Ancient Science Through the Golden Age of Greece* (New York: Dover, reprint 1993).

Shirley, J. W. *Thomas Harriot, A Biography* (Oxford: Clarendon Press, 1983).

Taton, R., and Wilson, C., eds. *Planetary Astronomy from the Renaissance to the Rise of Astrophysics. Part A: Tycho Brahe to Newton* (Cambridge: Cambridge University Press, 1989). Volume 2A of *The General History of Astronomy* series.

Taton, R., and Wilson, C., eds. *Planetary Astronomy from the Renaissance to the Rise of Astrophysics. Part B: The Eighteenth and Nineteenth Centuries* (Cambridge: Cambridge University Press, 1995). Volume 2B of *The General History of Astronomy* series.

Thoren, V. E. *The Lord of Uraniborg, a Biography of Tycho Brahe* (Cambridge: Cambridge University Press, 1990).

Tombaugh, C. W., and Moore, P. *Out of the Darkness, The Planet Pluto* (London: Lutterworth, 1980).

Turner, H. H. *Astronomical Discovery* (London: Edward Arnold, 1904).

Van Helden, A. *Measuring the Universe, Cosmic Dimensions from Aristarchus to Halley* (Chicago and London: The University of Chicago Press, 1985).

Westfall, R. S. *Never at Rest, A Biography of Isaac Newton* (Cambridge: Cambridge University Press, 1980).

Woolf, H. *The Transits of Venus, a Study of Eighteenth Century Science* (Princeton, New Jersey: Princeton University Press, 1959; New York: Arno Press, reprint 1981).

ARTICLES, MONOGRAPHS, AND PAMPHLETS

Adams, J. C. "Address Delivered by the President, Professor Adams, on Presenting the Gold Medal of the Society to M. Le Verrier." *Monthly Notices of the Royal Astronomical Society* **36**, 232–246 (1876).

Anonymous, "Suspected Existence of a Zone of Asteroids Revolving between Mercury and the Sun. Substance of M. Le Verrier's Letter to M. Faye," *Monthly Notices of the Royal Astronomical Society* **20**, 24–26 (1860).

Anonymous, "Supposed intra-Mercurial planet," *American Journal of Science,* Series 2, **29**, 296 (1860).

Anonymous, "Edward Claudius Herrick," *Obituary Record of Graduates of Yale College: Deceased from July 1859 to July 1870.* Presented at the Annual Meeting of the Alumni, 1860–1870; See also *American Journal of Science* Series 2, **34**, 159–160 (1862).

Anonymous, "A Supposed New Interior Planet," *Monthly Notices of the Royal Astronomical Society* **XX**, 98–100 (1860).

Anonymous, "The Planet Vulcan," *The Photographic News* April 13, 169 (1877).

Ashbrook, J., "Julius Schmidt and the Moon," *Sky & Telescope* **17**, 290–291 (1958).

Babinet, J., "Mémoire sur les nuages ignés du Soleil considérés comme des masses planétaires," *Comptes Rendus de l'Académie des Sciences* **22**, 281–286 (1846).

Backhouse, T. W., "Search for Vulcan," *The Journal of the Liverpool Astronomical Society* **4**, Pt. VI, 41–42 (1886).

Bates, R., and McKelvey, B., "Lewis Swift, The Rochester Astronomer," *Rochester History* **IX**, No. 1 (January, 1947).

Baum, R. M., "Le Verrier and the Lost Planet," *1982 Yearbook of Astronomy* (London: Sidgwick & Jackson, 1981), pp. 150–161.

Benzenberg, J. F., "Die Sternschnuppen sind Steine aus den Mondvulkanen," *Ueber die dunkeln Körper, die man zuweilen vor der Sonne hergehen sieht.* (Bonn, 1834), pp. 45–47.

Bless, R. C., *Washburn Observatory 1878–1978* (University of Wisconsin-Madison, 1978).

Brewster, D., "Recent Discoveries in Astronomy," *North British Review,* **XXXIII**, (65), 1–20 (1860).

Buys-Ballot, C. H. D., "Über den Einfluss der Rotation der Sonne auf die Temperatur unserer Atmosphäre." *Annln. Phys. Chem.* **68**, 205–213 (1846).

Buys-Ballot, C. H. D., "Lettre de M. Buys Ballot à M. Le Verrier," *Comptes Rendus de l'Académie des Sciences* **49**, 812–813 (1859).

Campbell, W. W., and Perrine, C. D., "The Lick Observatory-Crocker Eclipse Expedition to Spain," *Publications of the Astronomical Society of the Pacific* **18**, 13–36 (1906).

Campbell, W. W., "The Closing of a Famous Astronomical Problem," *Popular Science Monthly* 494–503 (1909).

Campbell, W. W., "Biographical Memoir of Edward Singleton Holden 1846–1914." *Biographical Memoirs, National Academy of Sciences* part of Vol. VIII (Washington DC, 1916).

Campbell, W. W., and Trumpler, R., "Search for Intramercurial Planets." *Publications of the Astronomical Society of the Pacific* **35**, 214–216 (1923).

Carrington, R. C. "On Some Previous Observations of Supposed Planetary Bodies." *Monthly Notices of the Royal Astronomical Society* **20**, 192–194 (1860).

Carrington, R. C., "Further Note on the Supposed Observation of an Intra-Mercurial Planet on the 12th of February, 1820," *Monthly Notices of the Royal Astronomical Society* **22**, 276 (1862).

Challis, J., "An Account of Observations Undertaken in Search of the Planet Discovered at Berlin on Sept. 23, 1846," *Memoirs of the Royal Astronomical Society* **16**, 415–426 (1847).

Challis, J., "On the Planet within the Orbit of Mercury, Discovered by M. Lescarbault," *Proceedings of the Cambridge Philosophical Society* **1**, 219–222 (1861); Reprinted in *Philosophical Magazine* **4**, 470–473 (1861).

Chapman, A., "Private Research and Public Duty: George Biddell Airy and the Search for Neptune," *Journal for the History of Astronomy xix*, 121–139 (1988).

Charlier, C. V. L., "Das Bodesche Gesetz und die sogenannten intramerkuriellen Planeten," *Astronomische Nachrichten* **193**, 269–272 (1913).

Clairaut, A. C., "Memoire sur la comete de 1682," *Journal des Sçavans* 38–45 (1759).

Comstock, G. C., "James Craig Watson, 1838–1880," *Biographical Memoirs of the National Academy* (Washington, DC: National Academy of Sciences, 1895), Vol. 3, pp. 43–57.

Comstock, G. C., "The Washburn Observatory," *Publications of the Astronomical Society of the Pacific* **9**, 31–33 (1897).

Curtis, H. D., "James Craig Watson, 1838–1880" *Michigan Alumnus* **44** (19), 306–313 (July, 1938).

Davidson, G., "Intra-Mercurial Planets," *The Sidereal Messenger* **3**, 113–115 (1884).

Denning, W. F., "A Supposed New Planet," *Science for All* **4**, 264–270 (1893).

Denning, W. F., "Search for an Intra-Mercurial Planet," *Knowledge* **23**, 134 (1900).

Dreyer, J. L. E., "Supposed Intra-Mercurial Planets," *The Observatory* **3**, 656 (1880).

Dunkin, E., "Obituary Notice of Johann Friedrich Julius Schmidt," *Monthly Notices of the Royal Astronomical Society* **45**, 211–218 (1885).

Eddy, J. A., "The Schaeberle 40-ft Eclipse Camera of Lick Observatory," *Journal for the History of Astronomy* **ii**, 1–22 (1971).

Eddy, J. A., "Thomas A. Edison and Infra-red Astronomy," *Journal for the History of Astronomy* **iii**, 165–187 (1972).

Eggen, O. J., "Vulcan," *Astronomical Society of the Pacific Leaflet* **287** (1953).

Einstein, A., Zur allgemeinen Relativitätstheorie; Zur allgemeinen Relativitätstheorie (Nachtrag); Erklärung der Perihelbewegung des Merkur aus der allgemeinen Relativitätstheorie; Die Feldgleichungen der Gravitation. *Sitzungsber. k. preuss. Akad. Wiss.* Pt.2 pp. 778–786, 799–801, 831–839, 844–847 (1915).

Faye, H., "Remarques de M. Faye à l'occasion de la Lettre de M Le Verrier," *Comptes Rendus de l'Académie des Sciences* **49**, 383–385 (1859).

Flammarion, C., "The Intra-Mercurial Planets," *The Popular Science Monthly* **14**, 714–721 (1879).

Fontenrose, R., "In Search of Vulcan," *Journal for the History of Astronomy* **iv**, 145–158 (1973).

Gould, B. A. Jr., *Report on the History of the Discovery of Neptune* (Washington DC: The Smithsonian Institution, 1850).

Gould, B. A. Jr., "New Planet, Inferior to Mercury," *The Astronomical Journal* **6**, 88 (1861).

Gould, B. A. Jr., "Aus einem Schreiben des Herrn. Dr. Gould an den Herausgeber," *Astronomische Nachrichten* **74**, 375–376 (1869).

Gregg, I., "The Planet of Romance," *Journal and Transactions of the Leeds Astronomical Society* (14) 16–28 (1906).

Grosser, M., "The Search for a Planet Beyond Neptune," in *Science in America Since 1820*, ed. by N. Reingold (New York: Science History Publications, 1976), pp. 303–323.

Hall, I. H., "Christian Henry Frederick Peters," Memorial Address by Isaac H. Hall, Ph. D., L. H. D. Metropolitan Museum of Art, New York, 1890.

Hanson, N. R., "Leverrier: The Zenith and Nadir of Newtonian Mechanics," *Isis* **53** (3), 359–378 (1962).

Herrick, E., "Lettre de M. Herrick à M. Le Verrier," *Comptes Rendus de l'Académie des Sciences* **49**, 810–812 (1859).

Herrick, E., "Supposed Planet between Mercury and the Sun," *American Journal of Science* Series 2, **28**, 445–446 (1859).

Hind, J. R., "Note on a dark, circular Spot upon the Sun's Disk, with rapid motion, as observed by W. Lummis, Esq., of Manchester, 1862, March 20," *Monthly Notices of the Royal Astronomical Society* **22**, 232 (1862).

Hind, J. R., "Obituary Notice of Urbain Jean Joseph Le Verrier," *Monthly Notices of the Royal Astronomical Society* **38**, 155–166 (1878).

Hind, J. R., "The Intra-Mercurial Planet or Planets," *Nature* **14**, 469–470 (1876).

Hind, J. R., "Stellar Objects seen during the Eclipse of 1869," *Nature* **18**, 663–664 (1878).

Holden, E. C., "Report of the Eclipse Expedition to Caroline Island May 1883," *Memoirs of the National Academy of Sciences* **2**, 5–146 (1884).

Janssen, J., "Note sur les passages des corps hypothétiques intra-mercuriels sur le Soleil," *Comptes Rendus de l'Académie des Sciences* **83**, 650–655 (1876); Abstract in *Nature* **14**, 534 (1876).

Johnson, M. J., "Address delivered by the President, M. J. Johnson, Esq., on presenting the Medal of the Society to M. Schwabe," *Monthly Notices of the Royal Astronomical Society* **17**, 126–132 (1857).

Kirkwood, D., "A world with a year of seventeen days," *Utica Morning Herald* May 7, 1873, p. 4, col. 3.

Kirkwood, D., "The Planet Vulcan," *The Popular Science Monthly* **13**, 732–735 (1878).

Langley, S. P., and Abbot, C. G., "A Preliminary Report of the Smithsonian Astrophysical Observatory Eclipse Expedition of May 1900," *Report of the Smithsonian Astrophysical Observatory 1891–1901, to the 57th Congress of the United States of America (Document #20, Exhibit D)* (Washington DC: United States Government Printing Office, 1902), pp. 295–308.

Ledger, E. E., *Intramercurial Planets: A Lecture Delivered in Gresham College, London on February 14, 1879 With an Appendix Containing Copies of Documents Referring to an Observation in the Year 1847* (Cambridge: Cambridge University Press, 1879).

Lescarbault, E. M., "Passage d'une planète sur le disque du soleil, observé à Orgères (Eure-et-Loir), par M. Lescarbault; Lettre à M. Le Verrier," *Comptes Rendus de l'Académie des Sciences* **50**, 40–46 (1860); Reprinted in *Annales de l'Observatoire Impérial de Paris Mémoires* **5**, 394–399 (1859): *Bulletin Internationale* published from the Paris observatory, January 3, 4, 5, 6, and 7, 1860, and *Cosmos* **16**, 50–56 (1860).

Lescarbault, E. M., "Observation d'une étoile d'un éclat comparable à celui de Régulus et située dans la même constellation. Extraite d'une Lettre de M. Edm. Lescarbault à M. le Secrétaire perpétuel," *Comptes Rendu de l'Académie des Sciences* **112**, 152 (1891); **112**, 260 (1891).

Le Verrier, U. J. J., "Détermination nouvelle de l'orbite de Mercure et de ses perturbations," *Comptes Rendus de l'Académie des Sciences* **16**, 1054–1065 (1843).

Le Verrier, U. J. J., "Discussion d'anciennes observations de Mercure, extraites par Édouard Biot de la Collection des vingt-quatre historiens de la Chine," *Comptes Rendus de l'Académie des Sciences* **17**, 732–737 (1843).

Le Verrier, U. J. J., "Nouvelles recherches sur les mouvements des planètes (premier Mémoire)," *Comptes Rendus de l'Académie des Sciences* **29**, 1–5 (1849).

Le Verrier, U. J. J., "Théorie et Tables du mouvement de Mercure. Chapt. XV, Recherches astronomique," *Annales de l'Observatoire Impérial de Paris* **5**, 1–196 (1859).

Le Verrier, U. J. J., "Lettre de M. Le Verrier à M. Faye sur la théorie de Mercure et sur le mouvement du périhélie de cette planète," *Comptes Rendus de l'Académie des Sciences* **49**, 379–383 (1859).

Le Verrier, U. J. J., "Examen des observations qu'on à présenteés, à diverse époques, comme pouvant appartenir aux passages d'une planète intra-mercurielle devant le disque du soleil," *Comptes Rendus de l'Académie des Sciences* **83**, 583–589; 621–624; 647–650; 719–723 (1876).

Le Verrier, U. J. J., "Note sur les planètes intra-mercurielles," *Comptes Rendus de l'Académie des Sciences* **83**, 561–563 (1876).

Liais, Emm, "L'absence de planètes prés du soleil," in *L'Espace Céleste et la Nature Tropicale Description Physique de l'Univers* (Paris: Garnier Frères, 1866), pp. 497–509.

Liais, Emm, "Sur la nouvelle planète annoncée par M. Lescarbault," *Astronomische Nachrichten* **52**, 369–378 (1860).

Liais, Emm, "Lettre de M. Liais," *Cosmos* **17**, 402–405 (1860).

Liais, Emm, "Sur d'anciens déplacements de taches sur le soleil à l'occasion de la note de M. Wolf, imprimée dans le compte rendu du 5. Mars 1860," *Astronomische Nachrichten* **54**, 139–144 (1860).

Liouville, J., "Sur un cas particulier du problème des trois corps," *Comptes Rendus de l'Académie des Sciences* **14**, 503–506 (1842).

Lockyer, N., "The Intra-Mercurial Planet Question," *Nature* **14**, 507 (1876).

Lockyer, N., "The Eclipse," *Nature* **18**, 457–462 (1878).

Lofft, C., "On the Appearance of an Opaque Body Traversing the Sun's Disc," *Monthly Magazine* 102–103 (March 1, 1818).

Main, R., "Sun-spots Suspected to be Identical with an Inter-Mercurial Planet," *Nature* **14**, 473 (1876).

Moigno, L'Abbé, "Lettre de M. Le Verrier à M. Faye sur la théorie de Mercure et sur le mouvement du périhélie de cette planète," *Cosmos* **15**, 471–476 (1859).

Moigno, L'Abbé, "Découverte d'une nouvelle planète entre Mercure et le soleil," *Cosmos* **16**, 22–28 (1860).

Moreux, Abbé Th., "Mercury and Intramercurial Planets," *Scientific American Supplement* **1680**, 163 (March 14, 1908).

Neumann, C., "Das Fernrohr für Sonnenbeobachtungen," *Wochenschrift für Astronomie, Meteorologie und Geographie* (14) 105–109 (April 3, 1861).

Nevalainen, J., "The Accuracy of the Ecliptic Longitude in Ptolmey's Mercury Model," *Journal for the History of Astronomy* **xxvii**, 147–160 (1996).

Newcomb, S., "On the Supposed Intra-Mercurial Planets," *The Astronomical Journal* **6**, 162–163 (1860).

Newcomb, S., "A Proposed Arrangement for Observing the Corona, and Searching for Intra-Mercurial Planets during a Total Eclipse of the Sun," *American Journal of Science and Arts* **47**, 413–415 (1869).

Newcomb, S., "Discussion and Results of Observations on Transits of Mercury from 1677 to 1881," *Astronomical Papers American Ephemeris and Nautical Almanac* **1**, 367–487 (1882).

Numbers, R. L., "The American Kepler: Daniel Kirkwood and His Analogy," *Journal for the History of Astronomy* **iv**, 13–21 (1973).

Oppolzer, Th. von., "Elemente des Vulkan," *Astronomische Nachrichten* **94**, 97–100 (1879).

Parkhurst, H. M., "Note on Interior Planets," *The Astronomical Journal* **6**, 142 (1861).

Perrine, C. D., "The Lick Observatory-Crocker Expedition to Observe the Total Solar Eclipse of 1901, May 17-18," *Publications of the Astronomical Society of the Pacific* **13**, 187–204 (1901).

Perrine, C. D., "Results of the Search for an Intramercurial Planet at the Total Solar Eclipse of August 30, 1905," *Lick Observatory Bulletin* **115**, 115–117 (1907).

Perrine, C. D., "The Search for Intramercurial Bodies at the Total Solar Eclipse of January 3, 1908," *Lick Observatory Bulletin* **5**, 95–97 (1909).

Peters, C. H. F., "That Intra-Mercurial Planet," Letter to the *Utica Morning Herald,* 1 (May 16, 1873).

Peters, C. H. F., "Some critical remarks on so-called intra-mercurial planet observations," *Astronomische Nachrichten* **94**, 321–336; 337–340 (1879). See also "The Intra-Mercurial Planet Question," *Nature* **20**, 597–598 (1879).

Pickering, E. C., "A Photographic Search for an Intermercurial Planet," *Scientific American* 154 (March 10, 1900).

Pickering, E. C., "A Photographic Search for an Intermercurial Planet," *Harvard College Observatory,* Circular No. 48 (1900).

Proctor, R. A., "The Planet Vulcan," *English Mechanic and World of Science* **24**, 160 (1876).

Pruett, J. H., "Icarus and the Case of Vulcan," *Sky and Telescope* 138–139 (April, 1950).

Radau, R., "Future Observations of the Supposed New Planet," *Monthly Notices of the Royal Astronomical Society* **20**, 195–197 (1860). See also "Faits des sciences," *Cosmos* **16**, 147–150 (1860).

Radau, R., "Réponse à M. Liais par M. Radau de Kœnigsberg," *Cosmos* **16**, 473–476 (1860).

Ranyard, A. C., "On a Remarkable Nebulous Spot Observed upon the Sun's Disc by Pastorff, May 26, 1828," *Monthly Notices of the Royal Astronomical Society* **34**, 26 (1873).

Ranyard, A. C., "Observations Made during Total Solar Eclipses," *Memoirs of the Royal Astronomical Society* (London: Royal Astronomical Society, 1879), Vol. XLI, Chapter XL.

Rodgers, J., "Letters Relating to the Discovery of Intra-Mercurial Planets," *Astronomische Nachrichten* **93**, 161–166 (1878).

Roseveare, N. T., "Leverrier to Einstein: A Review of the Mercury Problem," *Vistas in Astronomy* **23**, 165–171 (1979).

Schaffer, S., "Uranus and the Establishment of Herschel's Astronomy," *Journal for the History of Astronomy* **xii**, 11–26 (1981).

Schwabe, H., "Extract of a Letter from M. Schwabe to Mr. Carrington," *Monthly Notices of the Royal Astronomical Society* **17**, 241 (1857).

Sidebotham, J., "Note on an Observation of a Small Black Spot on the Sun's Disc," *Proceedings of the Literary and Philosophical Society of Manchester* **XII**, 105 (1873).

Smart, W. M., "On the Motion of the Perihelion of Mercury," *Monthly Notices of the Royal Astronomical Society* **82**, 12–19 (1921).

Smart, W. M., "John Couch Adams and the Discovery of Neptune," *Occasional Notes of the Royal Astronomical Society* (11) (August 1947).

Smith, R. W., "William Lassell and the Discovery of Neptune," *Journal for the History of Astronomy* **xiv**, 30–32 (1983).

Smith, R. W., "The Cambridge Network in Action: The Discovery of Neptune," *Isis* **80**, 395–422 (1989).

Smith, R. W., and Baum, R., "William Lassell and the Ring of Neptune: A Case Study in Instrumental Failure," *Journal for the History of Astronomy* **xv**, 1–17 (1984).

Swift, L., "Schreiben des Herrn. L. Swift an den Herausgeber," *Astronomische Nachrichten* **95**, 319–324 (1879).

Swift, L., "Intra-Mercurial Planets—One, Many or None?" *The Sidereal Messenger* **2**, 122–123 (1883-1884).

Swift, L., "The Intra-Mercurial Planet Question Not Settled," *The Sidereal Messenger* **2**, 242–244 (1883-1884).

Swift, L., "Intra-Mercurial Planets," *English Mechanic and World of Science* **64**, 135, Letter 39128 (1897).

Swift, L., "Report of Council to the Fortieth Annual General Meeting of the Royal Astronomical Society," *Monthly Notices of the Royal Astronomical Society* **20**, 147–148 (1860).

Thomson, W. (Lord Kelvin), "On the Mechanical Energies of the Solar System," *Transactions of the Royal Society of Edinburgh*, April (1854); reprinted in *Mathematical and Physical Papers* **II**, 1–27 (1884).

Tice, J. H., "The Supposed Planet Vulcan," *Scientific American*, (December 16, 1876), 389.

Tisserand, F., "Notice sur les planètes intra-mercurielles," *Annuaire de Bureau des Longitudes* 1882, 729–772.

Todd, D. P., "On the Use of the Electric Telegraph during Total Solar Eclipses," *Proceedings of the American Academy of Arts and Sciences* **16**, 359–363 (1881).

Todd, D. P., "On Observations of the Eclipse of 1887, Aug. 18, in Connection with the Electric Telegraph," *American Journal of Science* 2nd series, **33**, 226–228 (1887).

Trumpler, R., "Search for Small Planets at the Triangle Points of Mercury and the Sun," *Publications of the Astronomical Society of the Pacific* **35**, 313–318 (1923).

Tuttle, H. P., "Reminiscences of a Search for 'Vulcan' in 1860," *Popular Astronomy* **7**, 235–237 (1899).

"Reports on the Total Solar Eclipse of August 7, 1869," by Rear Admiral John Rodgers, *Astronomical and Meteorological Observations made at the United States Naval Observatory during the Year 1867*, Appendix II (Washington D.C., 1870).

"Reports of the Total Solar Eclipses of July 29, 1878 and January 11, 1880," *Astronomical and Meteorological Observations made during the Year 1876 at the United States Naval Observatory*. Part II, Appendix III (Washington D.C., 1880).

Van Helden, A., "The Importance of the Transit of Mercury of 1631," *Journal for the History of Astronomy* **7**, 1–10 (1976).

Watson, J. C., "Discovery of an Intra-Mercurial Planet," *American Journal of Science* 3rd Series, **16**, 230–233 (1878).

Watson, J. C., "On the Intra Mercurial Planets; from letters to the editors, dated Ann Arbor Sept. 3d, 5th and 17th, 1878," *American Journal of Science*, 3rd Series, **16**, 310–315 (1878).

Watson, J. C., "Sur l'existence d'une planète intra-mercurielle observée pendant l'eclipse totale de Soleil du 29 juillet. Lettre de M. J. Watson à M. Fizeau," *Comptes Rendus de l'Académie des Sciences* **87**, 376–377 (1878).

Watson, J. C., "Planètes intra-mercurielles observées pendant l'eclipse totale de Soleil du 29 juillet 1878. Lettre de M. Watson à M. Mouchez," *Comptes Rendus de l'Académie des Sciences* **87**, 786–788 (1878).

Weber, X., "Teleskopisches Meteor vor den Sonne, beobachtet am 5 Juli in Peckeloh," *Wochenschrift für Astronomie, Meteorologie und Geographie*. **N.F. 12**, 279–280 (1870).

Wilde, S., "Was it Vulcan?" *Scientific American* **35**, 304–305 (1876).

Wilson, C., "Clairaut's Calculation of the Eighteen-Century Return of Halley's Comet," *Journal for the History of Astronomy* **24**, 1–15 (1993).

Wolf, R., *Mittheilungen über die Sonnenflecken* **10**, 288–291 (1859). See also *Monthly Notices of the Royal Astronomical Society* **20**, 100–101 (1860).

Wolf, R., "Schreiben des Herrn. Professor Wolf an den Herausgeber," *Astronomische Nachrichten* **52**, 287–288 (1860).

Wolf, R., "Lettre de M. R. Wolf à M. Le Verrier," *Comptes Rendus de l'Académie des Sciences* **83**, 510 (1876).

Zachary, W. W., "An Historical Analysis of the Theoretical Solutions to the Problem of the Advance of the Perihelion of Mercury," Ph.D. thesis (University of Wisconsin: unpublished 1969).

OTHER PUBLICATIONS

Vulcan aroused great interest in the popular press of France, Great Britain and the United States of America. The most frequently consulted newspapers were the *London Times* and the *New York Times* for the years 1846, 1860, and 1878.

Index